Research, Action and Policy: Addressing the Gendered Impacts of Climate Change

Margaret Alston • Kerri Whittenbury
Editors

Research, Action and Policy: Addressing the Gendered Impacts of Climate Change

Springer

Editors
Prof. Margaret Alston OAM
Department of Social Work
Monash University
Victoria 3145, Caulfield East
Australia

Dr. Kerri Whittenbury
Department of Social Work
Monash University
Victoria 3145, Caulfield East
Australia

ISBN 978-94-007-5517-8 ISBN 978-94-007-5518-5 (eBook)
DOI 10.1007/978-94-007-5518-5
Springer Dordrecht Heidelberg New York London

Library of Congress Control Number: 2012950337

Printed on acid-free paper

Springer is part of Springer Science+Business Media (www.springer.com)

About the Book

This book is divided into five parts; the first is an introductory chapter by Margaret Alston discussing the significance of gender in the context of climate change. The second part on climate change and gender justice includes four chapters addressing the nature of knowledge and climate science. The first chapter by Nancy Tuana interrogates knowledge in the climate change space, the hidden value judgments and priorities inherent in climate discussions and proposed actions. Israel and Sachs' chapter further develops the notion of a feminist science, arguing there are several points where feminists can usefully intervene to challenge contemporary constructions of climate science. They note that scientific epistemology results in the separation of facts from values and politics, leading to an inherent failure to address the social justice issues implicit in global responses to climate change. Their call for feminist intervention at global levels is timely.

Bell takes the discussion on theoretical constructions further, arguing for the development of post-conventional theory to link gender, social justice and the environment. She criticises current approaches as being masculinist and imperialist. She draws on several feminist theorists including Vandana Shiva to argue for a theory of earth democracy incorporating a sustainable environment. The fourth chapter in this significant part is by well-known environmental activist Elaine Enarson, who argues for climate adaptation and disaster reduction efforts to come together in one 'big tent' rather than developing as 'two solitudes', or two varied disciplinary traditions. She urges the two to 'bridge' the gap drawing on the commonality of gender justice and environmental risk reduction.

Part III interrogates climate policy from a gender perspective. The first by Lena Dominelli notes the differential impacts of climate events on the poor and marginalized. In her chapter she focuses on gender relations in climate change and disasters, highlighting the neglect of women in climate policy. Her chapter draws on her research in Sri Lanka following the 2004 tsunami. She notes the gendered inequalities inherent in donor-related aid and advocates for a more transparent focus on gender in humanitarian aid situations. She notes that a failure to do this reduces women's rights and silences their claims.

Bee, Biermann and Tsachkert's chapter reinforces this call for gender-sensitive climate policy. They draw on a rights-based framework to highlight the need for social policies that address the needs of the most vulnerable. At the same time they resist the urge to view women as 'chief victims and caretakers' arguing that women's agency be recognised. They urge gender justice and positive social protection programs in order to address inequalities and promote structural change. Fletcher's chapter takes us to Canada and links the neoliberal notion of a free trade policy with its environmental inefficiencies and, ultimately, with the gendered impacts of these policies on Canadian farm women's lives. This chapter challenges policy from a different angle, noting the environmental and gendered consequences of a particular global policy framework. Jo Clarke's chapter moves us to Australia, where she critiques the farm exit policy adopted by the Australian government in response to climate events and reduced water. She notes the gendered consequences and uncertain outcomes of this policy.

Part IV addresses gendered climate change actions. Gotelind Alber, representing GenderCC, presents the findings of a campaign focused on gender and green energy use. She notes the motivations and barriers that challenge green energy adoption and makes recommendations on ways to further motivate adoption. Lorena Aguilar notes the work of the International Union for the Conservation of Nature (IUCN) in her discussion of gender inequalities in climate impacts and the need for the empowerment of women and the recognition of the work they undertake. She calls for gender-sensitive strategies to incorporate women and provides examples from across the globe. Rachel Goldsmith, Irina Feygina and John Jost discuss the way system justification can hamper efforts to adapt to climate change. They view this as a gender issue as men are more resistant to change, perhaps because they have more to lose. They raise the complex notion of psychological barriers to adoption of more sustainable practices and urge more attention to this limiting factor.

Part V draws on research from across the globe to note the gendered impacts of climate challenges. The first by Margaret Alston discusses the gendered impacts in Australia and the Pacific region. The second by Yianna Lambrou and Sibyl Nelson discusses outcomes in rural areas of India. The third by Kerri Whittenbury raises the issue of rising gender-based violence in the context of climate events. The fourth chapter in this part by Bettina Koelle discusses a project underway in South Africa engaged in a participatory action research project designed to empower women. Tatloghari and Paris discuss the issue of gender in the context of climate events in rice growing areas of the Philippines. Naomi Godden moves the discussion to South America, where she examines outcomes in fishing villages of Peru. Reetu Sogani brings us the voices of women from the Himalayan area where ice melt is causing significant consequences for women and men in these areas. Desley Hargraves brings us a significant understanding of the issues for Australian support workers who are part of the disaster response team working in Australia and the Asia-Pacific rim.

This book represents the voices of women from every continent and from vastly different climate events and challenges. There are commonalities in the voices of women and challenges to the global community to listen to and respond to women's experiences, to empower women, to provide gender-based supports and to ensure that women are part of the decision-making team addressing this major issue of our time. We urge readers to view this work as the beginning of a new way of understanding climate change that incorporates gender justice and human rights for all. We look forward to continuing our work in this space.

Margaret Alston OAM, Director, Gender, Leadership and Social Sustainability (GLASS) Research Unit, Monash University, Melbourne, Australia. Convenor, Gender and Climate Change: Women, Research and Action Conference.

Preface

This book emerges from the Gender and Climate Change: Women, Research and Action Conference held at Monash University Prato Campus, in Italy in 2011. Monash University's Gender Leadership and Social Sustainability (GLASS) Research Unit convened this conference which brought together researchers, policy makers and community advocates from across the globe to discuss the latest research findings in the area of gender and climate change and to develop strategies to address differential gendered impacts. GLASS academics led by the writer, senior research fellow, Kerri Whittenbury, and several Ph.D. students working through GLASS, participated in the conference. The timing of the conference was critical in that conveners and speakers hoped that information shared at the conference can provide input for significant international fora including Conferences of Parties (COP), Commission on Status of Women (CSW) and Earth Summit.

Delegates from 24 countries and from every continent discussed, debated and analysed how climate change differentially impacts women and men. Delegates also discussed the way research can inform policy makers and encourage action that recognises and addresses these gendered impacts and opportunities of climate change. To make the conference as constructive and progressive as possible there was a strong focus on developing shared priorities and platforms for further research, policy and advocacy work.

To enable widespread global input, GLASS received significant support from AusAid (ISS program) and from the US Department of State (Secretary's Office of Global Women's Issues) to sponsor delegates from the Asia-Pacific region, African nations and South America. A global representation was critical to facilitating a comprehensive discussion and reinforced that despite diversity, there are common experiences and similar ideas and priorities.

The conference included nearly 60 paper presentations across the three themes of:

- Gendered impacts of climate change
- Law, policy and frameworks
- Advocacy and innovation

The final afternoon of the conference was devoted to collaborative workshops to determine key priorities for research and action. The following outcomes statement is a summary of the workshops (GLASS 2011).

Delegates affirmed that climate change provides a unique opportunity for change – change that incorporates and facilitates gender equality and women's empowerment in programs, research, action and advocacy.

Critical issues requiring action that emerged from the conference are:

- ***That our understanding of climate change be reframed to incorporate analyses of gender and particular impacts on women***
- ***That gender be viewed as a critical factor in climate change analysis, policy and planning***
- ***That a technological/scientific focus alone does not address critical issues facing women and their families. The inclusion of a human rights, social justice and gender framework is essential for climate change analysis***
- ***That vulnerable countries and groups within countries be supported to positively adapt to climate change***
- ***That issues of food and water security encompass an analysis of impact on women that the global community mandate for reduced emissions***
- ***That green technologies be resourced and developed and the adoption of these at community levels be facilitated***
- ***That the development of green technologies provides a unique space for women's empowerment and involvement***
- ***That women be viewed not just as victims but as equal partners with men facing the challenges of climate change***
- ***That women be given equal representation on decision-making bodies and that these bodies address social and community impacts***
- ***That women's local knowledge be valued and incorporated in climate change actions and research***
- ***That research gaps be addressed in areas such as forced migration resulting from climate change; climate change and cities; managing technological solutions; renewable energies – and that this research have a gendered lens***
- ***That all research and policies include gender-disaggregated data and analysis***

About the Authors/Editors

Margaret Alston OAM Professor Margaret Alston, B. Soc. Stud (Syd), Dip. Comp. Applic. (RMIHE), M. Litt. (UNE), Ph.D. (UNSW), assumed duties as Head of Department of Social Work at Monash University in July 2008. She has established the Gender, Leadership and Social Sustainability (GLASS) research unit at Monash which now has 16 Ph.D. students and two post-doctoral appointees. Prior to commencing at Monash she was Professor of Social Work and Human Services and Director of the Centre for Rural Social Research at Charles Sturt University. She is an Honorary Professor at the University of Sydney and Charles Sturt University. She has served on a number of Boards including the Foundation for Australian Agricultural Women, Family Services Council, Family and Community Services Department in Canberra and the National Women's Advisory Group overseeing the Rural Women's Policy Unit in the Department of Primary Industries and Energy. She is Chair of the Australian Council of Heads of Schools of Social Work and is an advisor to the socio-economic working group for the Murray-Darling Basin Authority and the Health Workforce Australia working group on teaching. In 2008 she was appointed to the Australian delegation attending the commission for the Status of Women meeting in New York. In 2009, 2007 and 2003 she spent time as a visiting expert in the Gender Division of the United Nations Food and Agricultural Organisation in Rome. She has published widely in the field of rural gender and rural social issues. She has also acted as a gender expert for UN–Habitat in Kenya in 2009. She is currently researching the gendered impacts of climate change in Australia, the Pacific, Bangladesh and the Philippines. She has been a keynote speaker at a number of national and international conferences over the last several years and is sought out for media commentary on the rural social condition, and on climate change and gender issues.

She received her Medal of the Order of Australia in 2010 for services to social work and the advancement of women, particularly in rural areas.

Kerri Whittenbury Kerri Whittenbury, Ph.D., M. Litt., B.A., is Senior Research Fellow in the Gender, Leadership and Social Sustainability (GLASS) research unit, Monash University, Melbourne Australia. Kerri is a sociologist with research interests in gender, rural communities, rural and farm families, irrigation communities and gender and climate change. Her recent research projects include: social impacts of drought and declining water in the Murray-Darling Basin – funded by Australian Government and Monash University; impacts of declining water on women's health and well-being – funded by Monash University; and influences on irrigators' decisions in an environment of reduced water availability: the importance of subjective knowledge – funded by the Cooperative Research Centre for Irrigation Futures. She is currently researching gendered impacts of climate variability and climate events in Bangladesh, funded by a partnership between Oxfam Australia and Monash University.

Contents

Contributors

Lorena Aguilar IUCN (International Union for the Conservation of Nature), Gland, Switzerland

Gotelind Alber Sustainable Energy and Climate Policy, Berlin, Germany

Margaret Alston Department of Social Work, Monash University, Caulfield East, VIC, Australia

Beth Bee Josef Korbel School of International Development, University of Denver, Denver, CO, USA

Karen Bell School of Humanities and Social Sciences, Charles Sturt University, Wagga Wagga, NSW, Australia

Maureen Biermann Department of Geography, Pennsylvania State University, University Park, PA, USA

Josephine Clarke Monash University, Melbourne, Australia

Lena Dominelli Durham University, Durham, NC, UK

Elaine Enarson Independent scholar, Boulder, Colorado, USA

Irina Feygina New York University, New York, NY, USA

Amber J. Fletcher Canadian Plains Studies, University of Regina, Regina, SK, Canada

Naomi Joy Godden Monash University, Caulfield, VIC, Australia

Rachel E. Goldsmith Rush University Medical Center, Chicago, IL, USA

Desley Hargreaves Commonwealth Department of Human Services, Canberra, ACT, Australia

Andrei L. Israel Pennsylvania State University, University Park, PA, USA

John T. Jost New York University, New York, NY, USA

Bettina Koelle Free University of Berlin, Germany

University of Cape Town, Cape Town, South Africa

Indigo Development & Change, Nieuwoudtville, South Africa

Yianna Lambrou Food and Agriculture Organization of the UN, Rome, Italy

Sibyl Nelson Food and Agriculture Organization of the UN, Rome, Italy

Thelma R. Paris Social Sciences Division, International Rice Research Institute (IRRI) DAPO, Metro Manila, Philippines

Carolyn Sachs Pennsylvania State University, University Park, PA, USA

Reetu Sogani Community Awareness Centre Nainital, Uttarakhand, India

Gerlie T. Tatlonghari Social Work Department, Gender and Social Sustainability Unit (GLASS), Monash University, Melbourne, Australia

Petra Tschakert Department of Geography, Pennsylvania State University, University Park, PA, USA

Nancy Tuana Penn State University, University Park, PA, USA

Kerri Whittenbury Gender Leadership and Social Sustainability Research Unit, Monash University, Caulfield East, VIC, Australia

List of Figures

List of Tables

Part I
Introducing Gender and Climate Change

Chapter 1
Introducing Gender and Climate Change: Research, Policy and Action

Margaret Alston

Abstract In the context of a potentially global climate crisis, does gender matter? This book sets out to answer this question, drawing together research from across the globe to demonstrate that in areas as diverse as Australia, Canada, Africa, Asia and Europe, there is emerging evidence that gender does matter. It matters because the experiences of women and men during and after times of climate crisis are different – a difference based on cultural norms and practices, on work roles and access to resources, on safety and security and on different levels of vulnerability resulting from a combination of these factors. We are pleased to present this book, a product of our Gender and Climate Change conference held in Prato, Italy, in 2011. This conference was organized and auspiced by the Gender, Leadership and Social Sustainability (GLASS) research unit at Monash University, Australia. The focus of the book is on research, policy and action – what we know, what we need to know, how we might formally respond in order that support for the most vulnerable is forthcoming and that actions taken are transparent and focused on increasing the resilience of all affected by climate change.

Keywords Gender • Climate change • Action • Policy research • Security

In the context of a potentially global climate crisis, does gender matter? This book sets out to answer this question, drawing together research from across the globe to demonstrate that in areas as diverse as Australia, Canada, Africa, Asia and Europe, there is emerging evidence that gender does matter. It matters because the experiences of women and men during and after times of climate crisis are different – a difference based on cultural norms and practices, on work roles and

M. Alston, B.Soc., M.Litt., Ph.D. (✉)
Department of Social Work, Monash University, Dandenong Road 900, 3145 Caulfield East, VIC, Australia
e-mail: margaret.alston@monash.edu

M. Alston and K. Whittenbury (eds.), *Research, Action and Policy: Addressing the Gendered Impacts of Climate Change*, DOI 10.1007/978-94-007-5518-5_1,

access to resources, on safety and security and on different levels of vulnerability resulting from a combination of these factors. We are pleased to present this book, a product of our Gender and Climate Change conference held in Prato, Italy, in 2011. This conference was organized and auspiced by the Gender, Leadership and Social Sustainability (GLASS) research unit at Monash University, Australia. The focus of the book is on research, policy and action – what we know, what we need to know, how we might formally respond in order that support for the most vulnerable is forthcoming and that actions taken are transparent and focused on increasing the resilience of all affected by climate change.

Three decades ago very few would have recognized, understood or left unchallenged the emerging discussion of potentially permanent climate change. Yet, for much of the latter part of the twentieth century and into the new century scientists have been coming to grips with data suggesting human activities are having potentially devastating consequences on the global climate. Because of the major economic and geopolitical fallout, significant divisions have emerged at a number of levels including at a global level between developed and developing nations and at a community level over who is responsible and what actions should be taken.

Perhaps the most vitriolic division is between, firstly, those who are convinced that critical interventions are required to address our rapidly changing climate and, secondly, those who remain skeptical about the likelihood of ongoing and permanent climate change. This latter group includes those who view potential changes as part of normal cycles of weather changes and, therefore, as requiring no additional actions. The two opposing positions are sometimes pursued with a fervor bordering on zealotry – a variation of climate fundamentalism that divides skeptics and true believers into opposing, and sharply differentiated camps. It is hardly surprising then that there has been an over-emphasis on techno-scientific data to prove or disprove climate change and to develop scientific and technologically based solutions. This has resulted in far less attention being given to the social and community impacts of highly variable weather patterns and limited attention to gendered consequences.

Yet, regardless of the scientific veracity of the evidence, and the fundamentalism inherent in the debate, it is clear that climate variation is having major consequences for people across the world and, in particular, for the poorest of the poor, the majority of whom are women. Over the last three decades no continent has been spared its share of climate events and these have led to high levels of mortality and morbidity and major social dislocation.

The United Nations Secretary-General's High-level Panel on Global Sustainability (2012: 13) reports that global population will increase from seven billion to almost nine billion by 2040, and that by 2030, the world will need at least 50% more food, 45% more energy and 30% more water at a time when climate variability is creating uncertainty around food and water security. Further the number of undernourished people grew by 20 million between 2000 and 2008; 884 million people lack access to clean water; and 2.6 billion people do not have basic sanitation. This Panel (2012: 13) notes:

> The current global development model is unsustainable. We can no longer assume that our collective actions will not trigger tipping points as environmental thresholds are breached, risking irreversible damage to both ecosystems and human communities. At the same time, such thresholds should not be used to impose arbitrary growth ceilings on developing countries seeking to lift their people out of poverty. Indeed, if we fail to resolve the sustainable development dilemma, we run the risk of condemning up to 3 billion members of our human family to a life of endemic poverty. Neither of these outcomes is acceptable, and we must find a new way forward.

We must find a way forward and this must include an understanding of significant gendered consequences; that climate change has critical consequences for women and men; and that women are particularly vulnerable to poverty, insecurity and violence during and after climate events.

Climate Change

Climate change is defined as 'a change of climate which is attributed directly or indirectly to human activity that alters the composition of the global atmosphere and which is in addition to natural climate variability observed over comparable time periods' (UNFCC 2011). Climate change refers to the build up of greenhouse gases in the atmosphere as a result of human-induced activities, which causes temperature rises, melting of the ice caps and sea level rises. Climate change events include both slow-onset (or incremental) events such as droughts, desertification, temperature and sea level rises, and catastrophic or disaster events such as floods, bushfires, mudslides, cyclones, and storm surges. Global warming has major consequences for the planet and is impacting on agricultural production cycles through incremental and catastrophic climate events thus threatening food and water security.

The contestation over whether climate change is reversible has led to measures being enacted in various countries to reduce emissions. However global efforts to introduce binding targets on greenhouse emissions have to date been unsuccessful. Yet it appears that the window of opportunity to act is rapidly closing. Hence climate change and climate disasters have been referred to as 'the greatest moral challenge of our time' (Rudd 2007) because the need for action is evident but the process to gain global cooperation has been painfully slow.

Climate variability is not just an environmental problem; nor is it happening in a vacuum. As Terry (2009) notes, it is one of a number of trends including globalisation, rising world population, conflict, economic crisis and unpredictable policies that are leading to food and water insecurity across the globe. In 2008, Britain's Chief Scientific Adviser, Professor John Beddington, described food and water insecurity resulting from climate change as 'the elephant in the room' (Randerson 2008). Climate change must be viewed in a wider global context that includes additional factors that are increasing global uncertainty.

Global Climate Politics

International efforts to address climate change formally began in 1988 with the establishment of the Intergovernmental Panel on Climate Change (IPCC) – a joint collaboration between the World Meteorological Organisation (WMO) and the United Nations Environment Program (UNEP). Its mission is to assess 'the scientific, technical and socioeconomic information relevant for the risk of human-induced climate change' (IPCC 2012a, b:1). This body works with governments and experts to produce reports on the state of climate change. The United Nations General Assembly launched the first negotiations around climate change in 1990 in what has subsequently been referred to as the United Nations Framework Convention on Climate Change (UNFCCC). The first assessment report of the IPCC served as the basis for the negotiations within the UNFCCC. Optimistically this negotiation opened with a global target of 20% reduction of greenhouse gas emissions.

The UNFCCC was adopted in 1992 in preparation for the United Nations Environment and Development meeting in Rio de Janeiro, Brazil, (the Earth Summit) and officially came into force in 1994. Other outcomes of the Earth Summit, including the Agenda 21 and the Rio Declaration, had transparent attention to gender – it was only the UNFCCC, maintaining its scientific lens, which failed to adopt a strong gender focus (Hemmati and Rohr 2009).

The first annual conference of the parties (COP) to the UNFCCC was held in 1995 in Berlin. Ambitiously it was hoped that the meeting would strengthen global commitment to addressing climate change. At the third COP held in Kyoto in 1997, the Kyoto Protocol was adopted and signatories agreed to work to develop binding targets for the reduction of greenhouse gases (UNFCCC nd). At this point the target for reductions had been lowered from 20 to 5% and still caused significant conflict and no binding agreement.

At the 1997 meeting an international women's forum – Solidarity in the Greenhouse – was held and attracted 200 women from 25 countries. This resulted in a letter to the Chair of the UNFCCC asking that gender be recognized in negotiations. Subsequent COPs included minor events addressing gender but it wasn't until COP 13 that a major breakthrough occurred (Hemmati and Rohr 2009). COP 13, held in Bali, produced the Bali Action Plan (UNFCCC 2012) and it was at this meeting that feminists first expressed their public concerns about the lack of attention to gendered consequences. Disquiet had first surfaced at COP 9 in Milan when an informal meeting of concerned delegates formed a network to pursue gender matters. This group transformed into GenderCC –Women for Climate Justice and was internationally recognized at COP 13 in Bali when they produced several position papers addressing the likely gendered impacts of climate change. The slogan of GenderCC is 'no climate justice without gender justice' and the continued activism of GenderCC at subsequent COPs and through the production of informed reports (see for example Alber 2011) keeps gender in focus (Gender 2012).

Despite nearly 200 nations signing the Kyoto Protocol, subsequent COPs have struggled with finding ways to introduce binding agreements. This annual failure to achieve global cooperation culminated in a concerted effort at COP 15 held in Copenhagen in 2010. At this COP, 115 world leaders came together to develop a global binding agreement. More than 40,000 people attended this meeting hoping at last to see a fruitful outcome. However tensions emerged between developed nations and powerhouses and developing nations like China and India who argued they should be given some leeway to develop their industrial base before committing to binding reduction targets. Critically affected smaller nations such as the Pacific Island communities urged significant global reductions and their continued frustration at the lack of action. The outcome was a face saving Copenhagen Accord that was not binding, merely committing countries to reducing emissions and revisiting the issue in 2015.

Copenhagen failed to achieve the long-hoped for conclusion to the need for binding international targets to reduce greenhouse gas emissions. However there was agreement that the global community should act to ensure that global temperatures do not rise more than 2°C and that a review of this target would occur in 2015 (UNFCCC 2009). It appears that the global community cannot or will not attend to the need for reductions and therefore the future looks particularly grim. Adopting binding greenhouse emissions targets is essential to protecting people and communities across the planet. Lack of agreement, an increasing global population and threatened food security suggests that the twenty-first century will be an era of global tension and crisis relating to food and water security.

The IPCC continues to produce reports to inform the COP process. The first report in 1990 noted the likelihood of irreversible climate change and the need for action; the second in 1995 to inform the Kyoto Protocol noted that, on the balance of evidence, climate change is human-induced; the third in 2001 noted that climate change was contributing to sea level rises and melting of the ice caps; the fourth report in 2007 noted unequivocally that the earth is warming, that air and ocean temperatures are rising, that melting of the ice caps is widespread and that sea levels are rising; the fifth report is underway and will be released in 2014 (IPCC 2012a, b).

In recognition of its work, the IPCC was awarded the Nobel Peace prize in 2007 (shared with former US politician Al Gore)

> for their efforts to build up and disseminate greater knowledge about man-made climate change, and to lay the foundations for the measures that are needed to counteract such change (Nobel Prize 2007).

There is no doubt that the work of the IPCC has been instrumental in achieving global awareness of climate change and the need for action. The push by GenderCC and other organisations and individuals to ensure that the IPCC reports include attention to gender is evident in the fourth report and in preparatory work for the fifth report. It is critical that a gender focus be introduced to the global understanding of climate change, not only because of the innate need to introduce a human rights perspective, but also because the weight of evidence suggests that there are differential gendered impacts. Additionally it is critical because gender mainstreaming will improve the efficiency of any actions taken; and because attempts to achieve gender equality are further threatened by climate events (Skutsch 2002; Hemmati and Rohr 2009). Yet an additional disturbing element of the COP process is the small percentage of women from all countries who are heads or members of national delegations. Female membership hovers at between 15 and 25%. This suggests that the focus on women's experiences may not be fully recognized in global negotiations.

Despite increasing evidence that women are more vulnerable, it has become apparent that climate change policies, frameworks, discourses and solutions are rarely gender-sensitive. It is important then that gender is recognized as a factor in vulnerability to climate challenges; is highlighted in ongoing negotiations, research and actions and that women form a significant part of the decision making bodies.

Hemmati and Rohr (2009: 165) note

> the continued absence of gender issues in the climate-change debate is striking, both in negotiations and in the development of mechanisms, instruments, and measures.

Defining Gender

Gender refers to the socially constructed behaviours, customs and attitudes that shape what it is to be female and male in various cultural contexts (see for example Connell 2002). Gender is relational, in that it is shaped in the context of female and male interactions and is historically, socially, culturally and contextually bound. Connell (1995, 2002) notes that gender relations are continually contested in the intimate sphere she terms the 'gender regime', and that these negotiations are reinforced in the wider societal level of the 'gender order' – the laws, institutions and frameworks (for example the UNFCCC) that create uneven power relations between women and men. The rules and frameworks of the gender order heavily influence negotiations at the intimate gender regime level, making it difficult for women to achieve gender equality in both public and private spaces. Nevertheless, across time and space, gendered identities are constantly renegotiated and, as Hooper (2000: 60) notes, they involve power struggles between women and men where identities are 'consolidated and legitimated or downgraded and devalued', and where the relationship between masculinity and power is 'articulated and re-articulated'. In the vulnerable space following disasters, it appears that gender inequalities are being consolidated and legitimated in ways that reduce women's adaptive capacity.

Defining Vulnerability

Gender is a significant indicator of vulnerability to climate change – a term that commonly refers to the factors that combine to produce a lack of capacity to respond effectively, in this case to disasters. Those most vulnerable are characterised by gender, poverty, educational disadvantage, reduced service access, lack of employment options and other aspects of socio-economic disadvantage.

Vulnerability in the post-disaster space is increased by:

- loss of control over natural resources, including water;
- loss of access to and control over means of production, information, and decision-making;
- time poverty;
- increased energy needed to develop strategies and generate income;
- breakdown of educational and employment strategies;
- shortage of basic services including education, health, and employment;
- increased exposure to unsafe conditions, social isolation, and increased work hours;
- reduced ability to support children in education;
- less capacity for local organising;
- unsustainable over-production to ensure short term income; and
- vulnerability to natural disaster (e.g. flooding, drought) (Dankelman 2010).

Gender as a Critical Factor in Vulnerability to Climate Events

More recently, it has become apparent that gender is a critical factor shaping vulnerability to climate and landscape-induced disasters (Lane and McNaught 2009, FAO 2007; Neumayer and Pluemper 2007). For men this is evidenced by the interruption of their productive capacity in agriculture, for example their need to migrate away from their families to secure income, and in their loss of livelihood, productive assets and land. Women are much more likely to be living in poverty, are less likely to own land and resources to protect them in a post-disaster situation, and have less control over production and income, less education and training, less access to institutional support and information, less freedom of association, and fewer positions on decision-making bodies. Women are more constrained by their responsibilities for the aged and children, and during and after a climate event they are more likely to die and are more exposed to violence (Dankelman and Jansen 2010; Alston et al. 2010).

Lane and McNaught (2009: 71–72) note that before and during climate disasters women are more likely to be responsible for the practical preparation of the household, informing family members, storing food and water, and protecting family belongings. In both developing and developed countries, men are more likely to liaise with government officials, prepare the outsides of buildings, make decisions about evacuation and timing, manage water resources, distribute emergency relief and receive and disseminate early warnings to the community. Dankelman (2010: 59) argues that women have less access to resources that are essential to disaster preparedness, mitigation and rehabilitation, and that their workloads increase not only because men are more likely to migrate to look for work but because of a lack of energy sources, clean water, safe sanitation and health impacts. Heavy workloads often result in girls dropping out of school to assist.

Evidence from our own work indicates that men's mental health may suffer during prolonged period of drought (see for example Alston and Kent 2008). Women's increased morbidity and insecurity following disasters is demonstrated in a number of studies including our own in the Murray-Darling Basin area of Australia (Alston 2011), in the Pacific (Alston and Vize 2010) and Bangladesh (Alston et al. 2011). This work supports that of others including Neumayer and Pluemper's (2007) study of 20 years of disasters that indicates women are 14 times more likely to die in a catastrophic climate event; Enarson (2009, 2006) who notes the significant vulnerability of women to violent attack following Hurricane Katrina in the United States; and a report from an earthquake in Christchurch in February 2011 suggesting that the numbers of women seeking refuge from violence in the first month following the disaster nearly doubled (Lynch 2011). In vulnerable areas, including the Indian project where I contributed as an expert advisor, it has also been demonstrated that women are also more likely to be food insecure and living in poverty (Alber 2011; Lambrou and Nelson 2010).

Research from across the world notes the higher mortality and greater vulnerability of women to food and water insecurity, to lack of access to health services such as family planning following disasters, to violence and abuse, and to poor health resulting from climate change events. These include studies from Australia (Alston et al. 2010), India (Lambrou and Nelson 2010), the United States (Enarson 2009; Harris 2010); Vietnam (Neefjes and Valerie 2010); Nigeria (Odigie-Emmanuel 2010), South Africa (Vincent et al. 2010), Brazil (Corral 2010) and Colombia (Tovar-Restrepo 2010). These studies reveal not only that culture and social mores shape responses in particular circumstances, but that there are also disturbingly common global gendered trends regardless of country, developed or developing world status or type of climate catastrophe.

UN WomenWatch (2011: 1) notes that

> in many ... contexts, women are more vulnerable to the effects of climate change than men [and] are especially vulnerable when they are highly dependent on local natural resources for their livelihood. Those charged with the responsibility to secure water, food and fuel for cooking and heating face the greatest challenges [and] when coupled with unequal access to resources and to decision-making processes, limited mobility places women in rural areas in a position where they are disproportionately affected by climate change. It is thus important to identify gender-sensitive strategies to respond to the environmental and humanitarian crises caused by climate change.

Women do not lack agency in this space. They hold critical local knowledge that can enhance climate adaptations and assist the development of new technologies to address climate change in areas related to energy, water, food security, agriculture and fisheries, biodiversity services, health, and disaster risk management. Women's traditional knowledge and practices will add value the development of new technologies to address climate change (Lane and McNaught 2009).

Coping, Adaptation, Resilience and Transformational Change

Women's increased vulnerability following disasters may lead to the adoption of unsustainable coping strategies rather than long-term viable adaptation strategies, a response we have observed in our own work in Australia, India and Bangladesh. Coping is the ability to respond to, and avoid, harmful impacts whereas adaptation is the ability to genuinely transform structure, functioning and organisation to survive (Pelling 2011). Coping strategies are short-term, immediate actions taken to address a dire situation. They are generally formed under stress and may be unsustainable in the long-term (Taylor et al. 2010, sourced from Koelle and Shackleton 2011). Adaptation strategies are long-term sustainable actions, oriented towards livelihood security, using resources efficiently, aided by planning and reflection and involving institutional change.

Adaptation is dependent on the resilience of the people and communities affected; a concept that refers to the capacity of people and communities to absorb change in a positive way. The IPCC (2008: 880) defines resilience as 'the ability of a social or ecological system to absorb disturbances while retaining the same basic structure and ways of functioning, the capacity for self-organisation, and the capacity to adapt to stress and change.'

Moving from coping to adaptation strategies is highly dependent on how resilient people feel, how risky change appears, how safe traditional practices may feel and the institutional supports provided to people to move through what are effectively uncertain change processes. There are several barriers to adaptation including physical (such as the disaster event), financial (levels of poverty and costs of adaptation), cognitive (assessment of risk and trust in institutions), normative behaviour patterns (safety in doing what one has always done and unwillingness to deviate), and institutional governance and structure (institutional inequities, social inequalities, lack of information sharing and institutional inflexibility) – all of which are influenced by gender relations (Jones 2010; Koelle and Shackleton 2011; Pelling 2011; Adger et al. 2007; Tsachakert 2011).

Building resilience is not only essential to adaptation, it is also integral to transformative change, a process requiring input from institutional structures including government, and attention to structural inequalities and oppressive customs that limit gender equality. Pelling (2011) defines three stages of this process of transformative change as *building resilience*, *transitioning* from usual practices to new adaptations and *transformative change* where social contracts are renegotiated, the causes as well as the symptoms of vulnerability are addressed and actions taken.

There is much to be done to move the global community to transformative change. We require research on the social and gendered impacts as well as scientific/technological and economic outcomes. We require policies and actions that will create a stable, fairer and more robust world. We require activism and action. This book emerges from one such action – our 2011 conference.

References

Adger WN, Agrawala S, Mirza MMQ, Conde C, O'Brien K, Pulhin J, Pulwarty R, Smit B, Takahashi K (2007) Assessment of adaptation practices, options, constraints and capacity. In: Parry ML, Canziani OF, Palutikof JP, van der Linden PJ, Hanson CE (eds) Climate change 2007: impacts, adaptation and vulnerability. Contribution of working group II to the fourth assessment report of the intergovernmental panel on climate change. Cambridge University Press, Cambridge, UK, pp 717–743

Alber G (2011) Gender, cities and climate change: thematic report prepared for cities and climate change global report on human settlements. www.unhabitat.org/grhs/2011. Accessed 29 Nov 2011

Alston M (2011) Gender and climate change in Australia. J Sociol 47:53–70

Alston M, Kent J (2008) The big dry: exacerbating the link between rural masculinities and poor health outcomes for rural men. Aust J Sociol 44(2):133–147

Alston M, Kerri W, Alex H (2010) The social impacts of declining water availability and ongoing drought in the Murray-Darling Basin: Short Report. Monash University Department of Social Work, Melbourne

Alston M, Vize S (2010) Gender and climate change in the Pacific. Gender, Leadership and Social Sustainability (GLASS) research unit, Monash University, Melbourne

Alston M, Whittenbury K, Haynes A (2011) Gender and climate change in Bangladesh. Paper presented to the international rural sociology conference, Chania, Greece, August. Gender, Leadership and Social Sustainability (GLASS) research unit, Monash University, Melbourne

Connell R (1995) Masculinities. Allen & Unwin, St Leonards

Connell R (2002) Gender. Wiley, Sydney

Corral T (2010) Gender perspectives in adaptation strategies: the case of pintadas solar in the semi-arid region of Brazil. In: Dankelman I (ed) Gender and climate change: an introduction. Earthscan, London, pp 138–144

Dankelman I (2010) Introduction: exploring gender, environment and climate change. In: Dankelman I (ed) Gender and climate change: an introduction. Earthscan, London, pp 1–20

Dankelman I, Jansen W (2010) Gender, environment and climate change: understanding the linkages. In: Dankelman I (ed) Gender and climate change. Earthscan, London/Sterling, pp 21–54

Enarson E (2006) Women and girls last?: averting the second post-Katrina disaster. http://understandingkatrina.ssrc.org/Enarson/. Accessed 14 Mar 2011

Enarson E (2009) Women gender and disaster. Sage Publications, New York

Food and Agricultural Organisation (FAO) (2007) Gender and climate change: existing research and knowledge gaps. Gender and Population Division, FAO, Rome, Rome

Gender CC (2012) Homepage. http://www.gendercc.net/metanavigation/home.html. Accessed 5 Feb 2012

Gender, Leadership and Social Sustainability (GLASS) research unit (2011) Outcomes statement from the gender and climate change conference. http://www.med.monash.edu.au/glass/. Accessed 5 Feb 2012

Harris R (2010) Gender aspects of climate change in the US gulf coast region. In: Dankelman I (ed) Gender and climate change: an introduction. Earthscan, London, pp 152–157

Hemmati M, Rohr U (2009) Engendering the climate change negotiations: experiences, challenges and steps forward. Gend Dev 17(1):19–32

Hooper C (2000) Masculinities in transition: the case of globalisation. In: Marchand M, Sisson Runyan A (eds) Gender and global restructuring. Sightings, sites and resistances. Routledge, London, pp 59–73

http://unfccc.int/essential_background/library/items/3599.php?such = j&symbol = FCCC/CP/2007/6/Add.1#beg. Accessed 5 Feb 2012

Intergovernmental Panel on Climate Change (IPCC) (2008) Fourth assessment report, p 869. http://www.ipcc.ch/publications_and_data/ar4/syr/en/annexessglossary-r-z.html. Accessed 29 Nov 2011

Intergovernmental Panel on Climate Change (IPCC) (2012a) Fifth assessment report. http://www.ipcc.ch/. Accessed 5 Feb 2012

Intergovernmental Panel on Climate Change (IPCC) (2012b) IPCC history. http://www.ipccfacts.org/history.html. Accessed 5 Feb 2012

Jones L (2010) Overcoming social barriers to adaptation: background note. Overseas Development Institute. http://www.odi.org.uk/resources/download/4945.pdf. Accessed 16 May 2011

Koelle B, Shackleton S (2011) Presentation to the World Universities Network Limits to Adaptation group. Penn State University, Penn

Lambrou Y, Nelson S (2010) Farmers in a changing climate: food security in Andhra Pradesh, India. Food and Agricultural Organisation of the United Nations (FAO), Rome

Lane R, McNaught R (2009) Building gendered approaches to adaptation in the Pacific. Gend Dev 17(1, March):67–80

Lynch K (2011) Spike in domestic violence after Christchurch earthquake, National http://www.stuff.co.nz/national/christchurch-earthquake/4745720/Spike-in-domestic-violence-after-Christchurch-earthquake. Accessed 14 Mar 2011

Neefjes K, Nelson V (2010) Responding to climate change in Vietnam: opportunities for improving gender equality. In: Dankelman I (ed) Gender and climate change: an introduction. Earthscan, London, pp 107–114

Neumayer E, Pluemper T (2007) The gendered nature of natural disasters: the impact of catastrophic events on the gender gap in life expectancy 1981 – 2002. http://ssm.com/abstract = 874965

Nobel Prize (2007) Nobel peace prize 2007. http://www.nobelprize.org/nobel_prizes/peace/laureates/2007/. Accessed 5 Feb 2012

Odigie-Emmanuel O (2010) The gender impact of climate change in Nigeria. In: Dankelman I (ed) Gender and climate change: an introduction. Earthscan, London, pp 123–129

Pelling M (2011) Adaptation to climate change: from resilience to transformation. Routledge, London

Randerson J (2008) Food crisis will take hold before climate change warns chief scientist. The Guardian, March 7. http://www.guardian.co.uk/science/2008/mar/07/scienceofclimatechange.food. Accessed 5 Feb 2012

Rudd K (2007) Interview ABC. The World Today, April 27, 2010. http://www.abc.net.au/worldtoday/content/2010/s2883605.htm. Accessed 28 Nov 2011

Skutsch M (2002) Protocols, treaties, and action: the "climate change process" viewed through gender spectacles. Gend Dev 10(2):30–39

Taylor A, Harris K, Ehrhart C (2010) Adaptation: key terms *Tiempo* Issue 77 October, pp 10–13.http://www.tiempocyberclimate.org/portal/archive/pdf/tiempo77low.pdf. Accessed 16 May 2011

Terry G (2009, March) No climate justice without gender justice: an overview of the issues. Gend Dev 17(1):5–18

Tovar-Restrepo M (2010) Climate change and indigenous women in Colombia. In: Dankelman I (ed) Gender and climate change: an introduction. Earthscan, London, pp 145–151

Tsachakert P (2011) Presentation to the limits to adaptation working group, Penn State University team, May 17

UN WomenWatch (2011) Women, gender equality and climate change. http://www.un.org/womenwatch/feature/climate_change/downloads/Women_and_Climate_Change_Factsheet.pdf. Accessed 29 Nov 2011

United Nations Framework Convention on Climate Change (UNFCC) (2011) Full text of the convention. http://unfccc.int/essential_background/convention/background/items/2536.php. Accessed 29 Nov 2011

United Nations Framework on Climate Change Convention (UNFCCC) (2009) Copenhagen climate change conference 2009. http://unfccc.int/meetings/copenhagen_dec_2009/meeting/6295.php. Accessed 5 Feb 2012

United Nations Framework on Climate Change Convention (UNFCCC) (2012) Report of the conference of the parties on its thirteenth session, held in Bali from 3 to 15 December 2007. Addendum. Part two: action taken by the conference of the parties at its thirteenth session

United Nations Framework on Climate Change Convention (UNFCCC) (nd) Issues in the negotiating process: a brief history of the climate change process. http://unfccc.int/cop7/issues/briefhistory.html. Accessed 5 Feb 2012

United Nations Secretary-General's High-level Panel on Global Sustainability (2012) Resilient people, resilient planet: a future worth choosing. United Nations, New York. http://www.un.org/gsp/sites/default/files/attachments/GSPReport_unformatted_30Jan.pd. Accessed 5 Feb 2012

Vincent K, Cull T, Archer ERM (2010) Gendered vulnerability to climate change in Limpopo province, South Africa. In: Dankelman I (ed) Gender and climate change: an introduction. Earthscan, London, pp 130–137

Part II
Questioning Gender and Climate Justice

Chapter 2
Gendering Climate Knowledge for Justice: Catalyzing a New Research Agenda

Nancy Tuana

Abstract A key theme of feminist science studies theorists is the question of whose interests are served by the knowledge that mainstream science deems worthy of development, and whose interests are served by the knowledge projects that are overlooked or ignored. A central concern animating this analysis is thus whether we have the knowledge we need to ensure climate justice. The aim of this essay is to catalyze a new climate change research agenda designed to locate epistemic gaps and injustices, to reveal the circulations of power regarding what is known and what remains unknown, to render transparent the ways in which knowledge is framed, and to examine whose interests are served by our current knowledges and ignorances about anthropogenic climate change. My analysis includes not only scientific approaches to climate change, but also the gendering of knowledges and ignorances in the work of theorists studying the topic of gender and climate change.

Keywords Gender • Climate change • Feminist theory • Inequality • Vulnerability

Feminist theorizing in the past decades has revealed a wealth of knowledge that had previously been overlooked or marginalized. To select just two examples, consider the work of feminist economists who demonstrated the importance of appreciating and counting the contribution of unpaid labor to the economy, much, though not all, done by women (Braunstein and Staveren 2011; Folbre 1995; Himmelweit 1995; Waring 1990) or the work of feminist theorists who in their study of sexualized violence transformed our understanding of practices such as marital rape, date rape, and sexual harassment that had been "normalized" in such a way that the violence was minimized or denied (Fricker 2006; Scheman 1980).

N. Tuana (✉)
Department of Philosophy, Penn State University, University Park, PA, USA
e-mail: ntuana@psu.edu

M. Alston and K. Whittenbury (eds.), *Research, Action and Policy: Addressing the Gendered Impacts of Climate Change*, DOI 10.1007/978-94-007-5518-5_2,

Feminists working in the field of epistemology and science studies provide important methodological resources for better understanding knowledge production, and, in particular, the intersections between power and knowledge. Interests and power often result in the view that only certain methods lead to knowledge, as well as impacting and limiting the domain of knowledge that is actively pursued. Assumptions about who is capable of certain types of knowledge and the systematic marginalization and social bias against certain groups frequently results in the phenomena of epistemic injustice (Fricker 2007; Hoagland 2003) where the experiences and knowledges of such groups have been obscured and at times even rendered unintelligible. The development of feminist standpoint theory has offered insights into how biases common to scientific practitioners get embedded into the practice of science, yet remain invisible, held in place both by systematic ignorance and by a view of scientific objectivity as eschewing all subjectivity and thus immune, at least over time, to interest, politics, or prejudice (Haraway 1989; Harding 1991; Wylie 2002). While social biases about gender, class, or race are most likely to inform and limit scientific practice in the biological and social sciences (Fausto-Sterling 2000; Haraway 1997; Martin 1987) no field of science is immune from these impacts (Traweek 1988). For reasons such as these, feminist science studies theories have argued that scientific knowledge is limited and partial, requiring the insights of feminist and other liberatory theorists[1] to encourage less partial, more adequate knowledges (Haraway 1988; Harding 1991). Feminist science studies has revealed the illusion and the harm of the Western view of science as performing what Donna Haraway called the "God trick" of offering knowledge that is disinterested, value-free, emerges from no social location, remains independent of human interests or politics, and simply provides access to "the way the world is" (Haraway 1988, pp. 581–4). In its stead, feminist theorists have argued that "we need not—indeed, must not—choose between "good politics" and "good science"... for the former can at least sometimes produce the latter, and the latter, at least in some cases, requires the former" (Harding 2004, p. 30). The point, then, is not to eliminate politics or interests from science, but to understand which interests advance knowledge and for whom, and which obstruct knowledge, again for whom. In other words, we always have to ask *whose interests are served* by the knowledge that mainstream science believes it is important to develop, and whose interests are served by the knowledge projects that are overlooked or ignored.

This question is crucial in many domains, but it is particularly salient in the study of anthropogenic climate change, where what we know and do not know is inextricably interlinked with issues of justice. Our knowledges and our ignorances (Oreskes and Conway 2008; Tuana 2006) will have a huge impact on who will live and who will die, who will benefit and who will be harmed. A central concern animating my analysis is thus whether we have the knowledges we need to ensure climate justice. While this is a question far too complex to address in the space of

---

[1]I use this term to refer to all theorists working to identify and eradicate oppression, including race theorists, postcolonial theorists, etc.

one essay, my goal is *to catalyze a research agenda* designed to locate epistemic gaps and injustices, to reveal the circulations of power regarding what is known and what remains unknown, to render transparent the ways in which knowledge is framed, and to examine whose interests are served by our current knowledges and ignorances about anthropogenic climate change.

This essay is designed to add this relatively overlooked research agenda on the gendering of knowledges and ignorances to current work on gender and climate change. Over the past decade, there has been a growing recognition of the need to identify gender differentiated impacts of climate change. Studies of how gender roles and gendered division of labor, as well as underlying economic, social, and political factors, can result in gender differences in the impacts of climate change have become an important and growing aspect of the climate change literature (Alston 2010; Brody et al. 2008; Dankelman 2010; Denton 2002; Johnsson-Latham 2010; Terry 2009). Important scholarship has also begun to identify gendered differences in the perception of climate change impacts and risks (Slovic 1999; Flynn et al. 1994; Davidson and Freudenburg 1996). In addition, valuable analyses of gender representation in the climate policy domain have also provided insights regarding the question of gender representation in the political domain. While there is still much to be done on these topics, an understanding of the gendering of knowledges and ignorances has lagged far behind. This essay serves as a call to research to encourage the development of new knowledges in this important domain.

I provide two examples of knowledge-ignorance/power linkages. The first examines hidden value judgments embedded in current models of climate impacts. The second begins to tease out a problematic dualism at the heart of our views of the regions and peoples most impacted by global warming.

Value Judgments and Reasons for Concern

In an analysis of the value judgments of economic modeling in the context of climate change, particularly the comparison of the value of harms or benefits that occur at different times, or what is called "discounting," Julie Nelson argues that both the value judgment about the moral weight to be accorded to future generations and what counts as rigorous research:

> can be traced to a hypervaluation of detachment. By over-emphasizing characteristics of distance, individuality, autonomy, and abstraction within economic thought, the practices of orthodox neoclassical economics have become severely impoverished. Feminist insights into the roles of connection and detachment in the history and philosophy of science can help in analysis of this situation, and in shaping economic practices that can better help address the real ethical and practical questions humanity urgently needs to face (Nelson 2008, 442).

In a similar fashion, I will argue that this same focus on objectivity, detachment, and abstraction results in systematic bias in how climate science represents the harms of climate change.

There is little doubt that much attention in the climate change literature has been focused on the harms of climate change and the need to "prevent dangerous anthropogenic interference with the climate system" (UNFCCC 1992, Article 2). The United Nations Framework Convention on Climate Change, an international treaty established in 1992 at the United Nations Conference on Environment and Development and currently ratified by 195 parties, aims to stabilize greenhouse gas concentrations in such a way as to ensure that it is possible for "ecosystems to adapt naturally to climate change," to do all that is possible to guarantee that "food production is not threatened," and to allow "economic development to proceed in a sustainable manner" (UNFCCC, Article 2).

The Intergovernmental Panel on Climate Change (IPCC) was established to provide comprehensive assessments of the current state of knowledge about climate change, its potential environmental and socio-economic impacts, and possible options for adapting to the negative impacts of climate change and mitigation options aimed at reducing future harms. Authors of the IPCC Third Assessment Report worked to synthesize information on climate risks and vulnerabilities and to provide criteria for judging which vulnerabilities might be labeled "key." This study identified what was labeled the "five reasons for concern":

1. Risks to unique and threatened systems;
2. Risks of extreme weather events;
3. Distribution of impacts;
4. Aggregate impacts;
5. Risks of large scale discontinuities (McCarthy et al. 2001, Ch. 19).

To communicate the relationship between temperature increases and the severity of what they identified as the five key vulnerabilities, the authors created what is now called "the burning embers" diagram, which is reproduced in Fig. 2.1. The designation, "burning embers," was a reflection of the color scheme used to represent risks and impacts, which starts with yellow and increases to dark red. The figure was updated in 2008 (the version on the right hand side in Fig. 2.1) to incorporate new data collected in the IPCC fourth assessment report. The updated figure was based on observations of already existing impacts from increases in global mean temperatures, better understanding of the likelihood and magnitude of impacts from global mean temperatures, clearer identification of which regions are likely to be most affected, and evidence concerning possibilities of irreversible changes and large impacts on multiple-century time scales (Smith et al. 2009). What the comparison between the two representations clearly illustrates is that in each of the five "reasons for concern" categories, the temperature range that would likely avoid "dangerous anthropogenic interference" is getting smaller.

The figure provides a simple measure of the severity of harm easy for those not experienced in interpreting graphs or probability distribution figures to understand. Well-versed in reading red as signaling danger and yellow as a marker of caution, it takes little training to see how risks and negative impacts are correlated with increases in global mean temperature: the darker the color, the higher the danger.

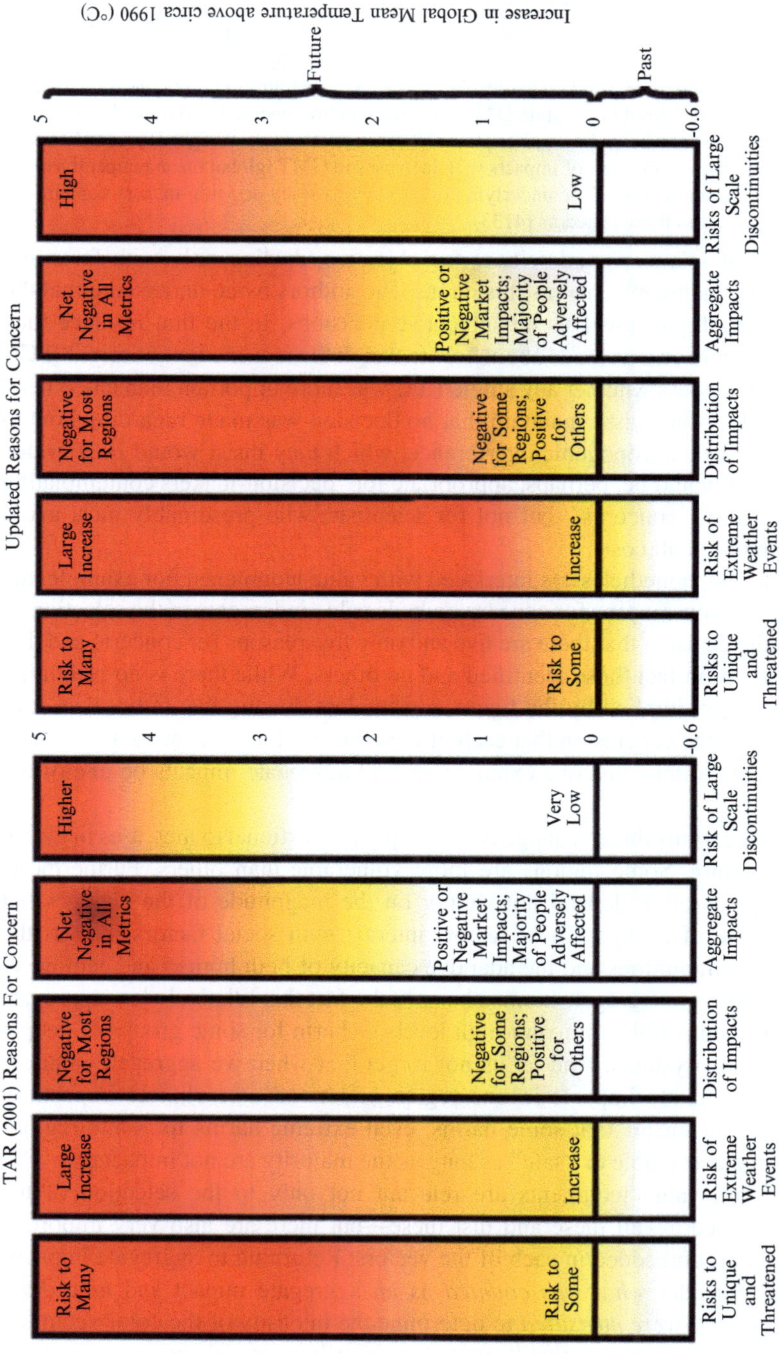

Fig. 2.1 Risks from climate change, by reason for concern on the left from climate change 2001: Impacts, adaptation and vulnerability compared with updated data from Smith et al. 2009. Climate change consequences are plotted against increases in global mean temperature (°C) after 1990

In their discussion of the update of the burning embers figure, Smith et al. (2009) underscore the objectivity of the IPCC's findings:

> In presenting the "embers" in the TAR [Third Assessment Report], IPCC authors did not assess whether any single RFC [reason for concern] was more important than any other; nor, as they noted, did they conclude what level of impact or what atmospheric concentrations of greenhouse gases would constitute DAI [dangerous anthropogenic interference], a value judgment that would be policy prescriptive. The "embers" were designed primarily to communicate the associations of impacts with increases in GMT [global mean temperature] and facilitate examination of the underlying evidence for use by decision-makers contemplating responses to these concerns (4133).

The view of science, good science, as objective, dealing only with facts and avoiding value judgments, haunts this quote. The authors twice underscore that the burning embers figure involves no subjective decisions. In the first instance they insist that no decision was made about how to weigh the reasons for concern: "IPCC authors did not assess whether any single RFC was more important than any other." In the second instance they maintain that no decision was made regarding how to define dangerous anthropogenic interference, which they insist would be a "value judgment" that would be, perhaps, appropriate for "decision-makers contemplating responses to these concerns," but not for scientists, who presumably must avoid value judgments at all costs.

But the figure, nonetheless, is interlaced with value judgments. For example, the selection of the five reasons for concern embeds value judgments in the selection of the five, in the decision that there are five and only five reasons for concern, and that the concerns are in fact those identified and no others. While there is no weighting across the vectors implied by the figure, nonetheless, having five rather than four vectors leads to the conclusion that each of these are vectors of concern to be taken seriously. Why, to take just one example, should aggregate impacts be one of the vectors?

The damage from climate change is not simply proportional to increases in global mean temperature. Some regions are more vulnerable than others. Furthermore, negative impacts often depend not simply on the magnitude of the changes but also on the rate of change. These factors interact with social factors such as the resilience of communities and the adaptive capacity of both human and ecological systems. A focus on aggregate harms obscures the fact that relatively low aggregate harms is still compatible with very high levels of harm for some groups of people and for some ecosystems. And let us not forget that when we aggregate impacts, the beneficial climate impacts in one region offsets adverse climate impacts in another. But that means that some harms, even extreme harms for some regions are considered acceptable or "safe" as long as the majority are not impacted.

Furthermore, value judgments are relevant not only to the selection of the reasons for concern—all these and just these—but there are also very important value judgments embedded in each of the vectors. Returning to aggregate impacts, we have to consider *what was counted* as an aggregate impact and *how those aggregate impacts were measured* to determine the intensity of the color vector on this figure. Were the impacts that "counted" primarily economic costs and perhaps loss of human life? What about those impacts that cannot be weighed via a simple economic model, such as loss of place (Tschakert et al. 2011; Tuana 2012) or loss of biological diversity (Thomas et al. 2004), or erosion of social and/or ecological resilience? Furthermore, second order value judgments can be embedded even after a first order value judgment is made. For example, a common first order value judgment is to count the loss of human life, or "years of life lost," as a key measure of aggregate impacts. But this is often accompanied by a second order value judgment, about how to count the value of life years. Are they all equivalent? Or do we use, as some have done, an economic measure that would result in lower values for the loss of life years in a poorer region of the world since individuals in these regions contribute a smaller amount to the GDP than individuals in wealthier regions (Dessai et al. 2004). Even rebalancing this inequity by weighting costs in poorer regions with income, embeds the value judgment that economic impacts (rather than psychological or aesthetic) are the appropriate measures. The decision to measure aggregate impacts in terms primarily of financial impact, frequently in terms of impact on gross domestic product, thus embeds many value decisions about what counts (income generating activities) and what does not (well-being, ecosystem flourishing), who counts, and who counts for less.

When we aim to "prevent dangerous anthropogenic interference," what is being counted? Is the measure geophysical thresholds, such as the disintegration of the West Antarctic Ice Sheet or the breakdown of the thermohaline circulation, or is the goal to calculate risks to social thresholds caused by water shortages or the food insecurities that collapse social orders or abilities of communities to adapt? Obviously, the geophysical thresholds are interlinked to social thresholds, but the bulk of the research to date has focused on the geophysical thresholds rather than the social thresholds.

To give just one final example, consider the first reason for concern. When the "risks to unique and threatened systems" figure is framed, what is counted as unique and what is overlooked in this mode of measurement? According to Smith et al. (2009, 4114) "This RFC [reason for concern] addresses the potential for increased damage to or irreversible loss of unique and threatened systems, such as coral reefs, tropical glaciers, endangered species, unique ecosystems, biodiversity hotspots, small island states, and indigenous communities." While some spatial regions are considered in this measurement (small island states), others are ignored (tropical ecosystems, dry land areas).

Joni Seager in "Death by Degrees" focuses on the ideological framing of the 2°C rhetoric. 2°C has been seen as an important threshold consensus achieved in the 2009 G8 Summit meetings, namely that in order to prevent dangerous anthropogenic interference with the climate system, the global average temperature should not be allowed to exceed 2°C over pre-industrial levels. Seager argues that this target is "distinctively ideological: it is precise enough to appear to be scientifically based, and it is a target for which geophysical systems themselves appear to be the primary reference (rather than emissions-control targets or historically based reduction goals, for example, which are directed towards human actions)" (Seager 2009, p. 14). She reveals the various components of this ideology including the view that humans can 'master' nature, or in this case, climate change and that warming up to 2° is "a degree of danger that is acceptable" (14), reminding us that "many ecosystems and peoples will hit limits to adaptation long before 2°C, and some already have" (15).

> For the millions of people in poor countries, low-latitude countries, low-lying states, and small island states, 2° is not acceptable. For the dozens of states already pushed to adaptive limits, a 2° cap, even if achievable, is too little, too late. For fragile ecosystems, perhaps especially coral reef and other marine communities, 2° of warming is not a safe target (2009, 16).

Aggregating harms and identifying "Reasons for Concern" is as distinctively ideological as the target of 2°C. Even worse, it is a trap and an illusion. It renders us complacent in believing that we are developing the knowledge we need to understand the harms of climate change, while at the same moment obscuring the actual harms done to individuals, to groups, and to ecosystems. Such models offer us value judgments obscured by a cloak of objective detachment, when what is needed for climate justice is value transparency, clear attention to all the impacts, and a science that cultivates a sentiment of responsibility and care instead of objective detachment. Unpacking what this would mean in practice in the domain of climate science is, I contend, a key research agenda for feminist scholars.

This layered example is designed to illustrate the important work needed by feminist scholars in the domain of climate change research. The image of science as objective and as both interest- and power-free frames our knowledges and ignorances about climate change impacts. It represents a form of the gender dimension of climate change that is imbedded in scientific practice and must be teased apart to fully address issues of climate justice. While this example focuses on the value judgments imbedded in scientific knowledges concerning climate impacts, in the next section my focus will be on values and assumptions embedded in the wider narratives of climate impacts, for they are framed by power/knowledge couplings as well.

The Feminist Dream of a Successor Science

In her study of the intersections between feminist and postcolonial science studies, *Science from Below,* Sandra Harding focuses her analysis on the knowledge projects and practices of Western modernity, arguing that modernity "remains haunted by anxieties about the feminine and the primitive, both of which are associated with the traditional" (2008, 1). She reminds us that any account of the development and expansion of Western science, as well as of Western society, must understand their deep connection to empires and colonization. The "development and modernization of the West were materially as well as ideologically *built on the exploitation,* de-development, and "constructed traditionalism" of the societies which European expansion encountered, from 1492 through the events in today's newspapers" (2008, 68). Furthermore, Harding details how Western scientific and technological development have gone hand-in-hand with widespread environmental destruction, the de-skilling and consequent alienation of labor, a steadily increasing gap between the resources available to rich and poor both locally and globally, and the continuing resources these provide for sexist and racist projects.

Harding claims that feminist and postcolonialist theorizing provide three important resources for understanding the complex interconnections between power and knowledge that are essential to practices of science that are more just. First, they reveal the processes and institutions that led to indigenous knowledge practices, other than those of Western science,[2] being ignored or suppressed. Second, through careful investigation of the co-constitution of gender with class, race, and other social hierarchies these forms of theorizing serve as an important reminder of the various others whose participation is essential to the emergence of science as a truly democratic practice. Third, they focus attention on women as active agents in the processes of scientific and technological change.

While these insights provide a powerful lens through which to render transparent the workings of power/knowledge-ignorance couplings in the study of the geophysical basis of climate change, my examples in this section turn instead on the social science narratives, particularly those dealing with impacts, adaptation, and vulnerability (the domain of Working Group II of the IPCC). Here, albeit briefly, I would like to add to our research agenda and provide an example of the ways in which the narratives surrounding global climate change are informed by and reinforce the framework Harding labeled "Western modernity." What Harding's work has so clearly revealed is that the interests that structure knowledge and determine what is known, as well as what is ignored, are those of the more powerful nations and those that matter to powerful organizations and corporations. She documents how science "from above" enacts a linked set of dualisms in which the first term is privileged. What I add to this insight is that this same dualism structures the dominant frameworks for thinking about global climate change as well (Fig. 2.2).

My focus in this section concerns how these same dualisms circulate in climate discourses to link agency with the "Global North" and vulnerability with the "Global South.[3]" And just as the dualisms are gender coded, I will argue that the discourses themselves carry gendered biases. Not only is the "Global South" viewed as vulnerable, but gendered tropes of this same dualism work to render women in

[2]As Harding reminds us, Western science, far from providing a "view from nowhere" is itself an indigenous knowledge practice embedded in and reinforced by traditional values and institutions.

[3]I fully realize that every term used to divide the world (First World/Third World—with or without the addition of the Second World; developed/developing; modern/traditional; high-income economies/low-income economies; etc.) are all fraught with problems. While I will use the designation of Global North/Global South (reflecting rankings on the Human Development Index), I do so throughout with quotes in recognition that distinctions like this embed values in ways not always fully transparent.

Fig. 2.2 Some of the central dualisms that structure Western modernity

Privileged	Subordinated
Objective and value free	Subjective and value laden
Science	Indigenous knowledges
Modern/Developed	Primitive/Less Developed
Knowledge	Folk wisdom/ignorance
Agency	Passivity
Masculine	Feminine

the "Global South" as victims. The "Global North" is depicted as having agency and assigned responsibility both for causing the problem of climate change, but also for responding to it with technological and policy solutions.

This dual trope of the "Global South" as victims of climate change in need of the resources of the "Global North" to survive is prevalent in climate science literature. Let me be clear, a justice perspective *does* require acknowledging the differential impacts of climate change on regions, as well as recognizing that some groups have more resources for adaptation than others. While not intending to deny either of these points, the recurring trope of conceptualizing the "Global South" as vulnerable and less able to act in the face of climate impacts, repeats and is informed by centuries of discourses regarding these countries as lessor—less developed, less modern, less technologically advanced, less stable, less capable of self-governance. *The problem is that while the rhetoric reflects certain truths, it plays into and perpetuates systematic prejudices about these countries embedded in the ontology of Western modernity.*

To give just one example, in the course of one page, Stephen H. Schneider and Janica Lane in their essay, "Dangers and Thresholds in Climate Change," mention the vulnerability of what they refer to as "the poorer, warmer nations of the world" nine times. Just a few illustrative quotes reveal the persistence of the above dualisms: developing nations "will experience more and more severe climate change impacts;" "the developing nations will most likely experience predominately detrimental effects;" "there is an imbalance between rich and poor nations' ability to cope with climate impacts;" "less developed countries tend to have lower adaptive capacities, as they are often limited by financial, technological, and governmental constraints;" "the uneven distribution of climate change impacts leaves the hotter, poorer nations—the countries that have less adaptive capacity—more vulnerable and more in need of adaptation" (2006, p. 28). Schneider's and Lane's aim is to argue that justice concerns must take these differences into account. Despite their good intentions, the same discourse that Harding warns us about haunts this text. The "less developed" are framed as less modern, less capable, less technologically advanced.

This same rhetorical repetition frames women in these countries. We are told that:

- Women constitute the largest percentage of the world's poorest people.
- Because of the "feminization of poverty," women are most likely to experience the greatest negative impacts of shifting weather patterns, resulting in further deprivations.
- Gender roles render women more vulnerable.
- Women have little voice in climate policy debates.

What is particularly worrisome is that these same tropes of lack and passivity inflect even feminist discourses. While calling for greater attention to the impact of climate change on women and for the greater participation of women in climate policy, the same rhetorical linkages between women-poverty-vulnerability circulate in feminist discourses.

An early essay on gender and climate change by Fatima Denton (2002) serves as illustration. We find multiple repetitions of the above themes throughout a paper that is calling for mainstreaming gender issues into debates on climate change and sustainable development, and the inclusion of women in decision-making. Consider the following quotes:

On the link between women and poverty:

> Climate change is likely to accentuate the gaps between the world's rich and poor. It is widely accepted that women in developing countries constitute one of the poorest and most disadvantaged groups in society (p. 11).

> Women are already paying huge prices for globalisation, economic depression, and environmental degradation. Climate change is likely to worsen their already precarious situation, and leave them even more vulnerable (p. 18).

On the interaction between poverty and climate harms:

> women and their livelihoods activities are particularly vulnerable to the risks posed by environmental depletion (p. 11).

On the relation between gender roles and women's vulnerability:

> poor women are generally on the receiving end of the effects of increasing environmental degradation and depletion of natural resources, because of their involvement in, and reliance on, livelihoods activities which depend directly on the natural environment (p.12).

> Gender inequalities continue to exist in terms of access to land, control over resources, ability to command and access paid labour, capacity, and strategies for income diversification, as well as time spent on agricultural or forestry-based activities (p. 17).

And on the absence of women from climate policy development:

> Women are patently absent from the climate change decision-making process (p. 11).

> Climate negotiations could be seen as a parody of an unequal world economy, in which men, and the bigger nations, get to define the basis on which they participate and contribute to the reduction of growing environmental problems, while women, and smaller and poorer countries, look in from the outside, with virtually no power to change or influence the scope of the discussions (p. 10).

> Power dynamics characterise the relationships between richer and poorer nations, and these have gendered implications. If poorer nations are finding it difficult to get richer nations to meet their obligations and work towards climate mitigation, poor women have an even bigger problem in promoting their agenda. If smaller and poorer nations have difficulties in mounting the necessary infrastructure to take advantage of CDM projects, poorer women have even fewer means and less scope to diversify their livelihood and look after their families (p. 17).

My point is that even in the context of arguing for gender differentiated impacts and solutions that will adequately address them, even feminist attention to gender and climate change can get caught in the logic of dualisms and *aggregate women*, or more specifically, women from the less developed nations, and the poor, and depict them, all of them, as more vulnerable.

While selecting only one text to reveal the pattern of this discourse to illustrate the women-poverty-vulnerability linkages for women in developing countries, these associations have been and continue to be prevalent in the literature (cf. Cannon 2002; Dankelman 2002; Demetriades and Esplen 2008; Hannan 2009; Nelson et al. 2002; Terry 2009). And as noted by Arora-Jonsson (2011) the trope of the North as agential has resulted in a corresponding image of Northern women as more environmentally virtuous, namely, "more sensitive to risk, more prepared for behavioural change and more likely to support drastic policies and measures on climate change."

Once again, we see the trope of Western modernity so clearly depicted by Harding (2008) at work in these texts. The strength of the linkages between the "Global South" and the "primitive" or, as we say, "less developed," and the related women[4]-poverty-vulnerability linkage have such a strong hold on our conceptual framework that we uncritically accept statements such as "70% of all poor people are women," as well as repeat and reinforce the message through labels like "the feminization of poverty." Sherilyn MacGregor, to cite just one example, in an important analysis of the need for research on "the ways in which gendered discourses, roles and identities shape the political and material aspects of climate change," unwittingly participates in this trope of repetition:

> There is widespread agreement among climate change analysts and policy makers that the more socially and economically marginalized people are, the more vulnerable they are to the effects of global warming. The poor will be hurt the most. However, few other than feminists put the global feminization of poverty into the frame. In his analysis, for example, Giddens (2009) refers to 'the poor' as a homogenous group, with no attention to the fact that women are more likely to be poor, and to be responsible for the care of poor children, than men. This is a problematic blindness. Approximately 70 per cent of the world's poor are women; rural women in developing countries are among the most disadvantaged groups on the planet. They are therefore unlikely to have the necessary resources to cope with the changes brought by climate change, and very likely to suffer a worsening of their everyday conditions (2010, 130).

[4]Specifically women from the "Global South" and aggregated to remove any differences.

Yet, as critiqued by Arora-Jonsson (2011), the generalizations about women and poverty are seldom questioned. "No scientific study is ever cited to document percentages such as the assertion that 70% of all poor people are women . . . there is in fact little gender disaggregated data to support the feminization of poverty thesis" (p. 746). An early paper by Cecile Jackson (1996) exposed a number of problematic assumptions linked to the feminization of poverty thesis. She takes issue with the World Bank assumption not only that women are disproportionately represented among the poor, but also their inference from this that "the poorer the family the more likely it is to be headed by a woman" (World Bank 1989, p. iv). Jackson notes that this linkage of women-headed household and poverty is not accurate. "Much depends on the reason for female headedness, those which are *de facto* household heads and receive remittances from migrant males may often be less poor than male-headed households" (1996, p. 492). She also notes that the relationship between female-headed households and income varies significantly across countries. Jackson laments that "the unfortunate term "the feminization of poverty" has come to mean not (as gender analysis would suggest) that poverty is a gendered experience, but that the poor are mostly women" (p. 491). Despite more than a decade of studies like Jackson's critiquing the women-poverty linkage, it continues to hold sway even in feminist literature.

Adding to the problem is the assumption that poverty and vulnerability go hand-in-hand. With women, or at least women from less developed countries, too often framed via the second term in the chart above, so that their vulnerability renders them passive, susceptible, acted upon. In this way, women's vulnerability becomes *the problem* of gender justice. But this then reinforces the maleness of agency, assumes that men are not vulnerable to climate change, makes it difficult to see or appreciate women's agency, and occludes the relevance of gender dynamics in the "Global North."

The importance of gender, just like the impacts of climate change, will not be effectively understood as long as we approach it from an aggregate perspective. Neither the poor nor women, even "women from the Global South," are homogeneous. As feminist scholarship has reminded us time and time again, while gender can serve as an analytic category, it must always be richly situated. Gender outside of its complex contexts is too rough a measure to be effective and will always risk repeating dominant patterns of thinking and thereby reinforce sexist assumptions about women. In short, gender must always be studied from specific and embodied locations where we as researchers are continually attentive to the complex and intersecting power relations from which gender emerges in its specific manifestations and where we as researchers are always "answerable for what we learn how to see" (Haraway 1988, p. 583). We must remember that our findings will always be "partial, locatable, critical knowledges sustaining the possibility of webs of connections called solidarity in politics and shared conversations in epistemology" (Haraway 1988, p. 584). It is in such practices and in our efforts to resist simplification, homogenization, and aggregation, that we feminists have a chance to achieve our dream of a successor science.

The slogan of the Gender and Climate Change Network (genderCC) is that "There will be no climate justice without gender justice." Achieving this important goal will require such a successor science and an expansion of what counts as research on gender and climate change to include attention to circulations of knowledge/ignorance and power.

References

Alston M (2010) Gender and climate change in Australia. J Sociol 47(1):53–70

Arora-Jonsson S (2011) Virtue and vulnerability: discourses on women, gender and climate change. Glob Environ Chang 21:744–751

Braunstein E, Staveren IV (2011) Embedding care and unpaid work in macroeconomic modeling: a structuralist approach. Fem Econ 17(4):37–41

Brody A, Demetriades J, Esplen E (2008) Gender and climate change: mapping the linkages a scoping study on knowledge and gaps, vol 44. BRIDGE, Institute of Development Studies, University of Sussex, Brighton

Cannon T (2002) Gender and climate hazards in Bangladesh. Gend Dev 10:45–50

Dankelman I (2002) Climate change: learning from gender analysis and women's experiences of organising for sustainable development. Gend Dev 10(2):21–29

Dankelman I (2010) Gender and climate change: an introduction. Earthscan, Washington, DC

Davidson DJ, Freudenburg WR (1996) Gender and environmental risk concerns: a review and analysis of available research. Environ Behav 28:302–339

Demetriades J, Esplen E (2008) The gender dimensions of poverty and climate change adaptation. IDS Bull 39(4):24–31

Denton F (2002) Climate change vulnerability, impacts, and adaptation: why does gender matter? Gen Dev 10(2):9–20

Dessai S, Adger WN, Hulme M, Turnpenny J, Köhler J, Warren R (2004) Defining and experiencing dangerous climate change. Clim Chang 64(1):11–25

Fausto-Sterling A (2000) Sexing the body: gender politics and the construction of sexuality. Basic Books, New York

Flynn J, Slovic P, Mertz CK (1994) Gender, race, and perception of environmental health risks. Risk Anal 14:1101–1108

Folbre N (1995) "Holding hands at midnight": the paradox of caring labor. Fem Econ 1(1):73–92

Fricker M (2006) Powerlessness and social interpretation. Epistem J Soc Epistemol 3(1):96–108

Fricker M (2007) Epistemic injustice: power and the ethics of knowing. Oxford University Press, Oxford

Giddens A (2009) The politics of climate change. Polity, London

Hannan C (2009) Gender mainstreaming climate change. Kvinder, Køn & Forskning NR (Women, Gender and Research) 3(3):46–51

Haraway D (1988) Situated knowledges: the science question in feminism and the privilege of partial perspective. Fem Stud 14(3):575–599

Haraway D (1989) Primate visions: gender, race, and nature in the world of modern science. Routledge, New York

Haraway D (1997) Modest_Witness@Second_Millenium.FemaleMan_Meets_OncoMouse. Routledge, New York

Harding S (1991) Whose science? whose knowledge? Thinking from women's lives. Cornell University Press, Ithaca

Harding S (2004) A socially relevant philosophy of science? Resources from standpoint theory's controversiality. Hypatia 19(1):25–47

Harding S (2008) Sciences from below: feminisms, postcolonalities, and modernities. Duke University Press, Durham
Himmelweit S (1995) The discovery of "unpaid work": the social consequences of the expansion of "work". Fem Econ 1(2):1–19
Hoagland SL (2003) Practices of knowing: transcendence and denial of epistemic credibility, or engagement and transformation. Int Stud Philos XXXV(2):21–37
Jackson C (1996) Rescuing gender from the poverty trap. World Dev 24(3):489–504
Johnsson-Latham G (2010) Why more attention to gender and class can help combat climate change and poverty. In: Dankelman I (ed) Gender and climate change: an introduction. Earthscan, London/Sterling, pp 212–222
Martin E (1987) The woman in the body: a cultural analysis of reproduction. Beacon, Boston
McCarthy JJ, Canziani OF, Leary NA, Dokken DJ, White KS (2001) Climate change 2001: impacts adaptation and vulnerability. Contribution of working group II to the third assessment report of the intergovernmental panel on climate change. Cambridge University Press, Cambridge
Nelson JA (2008) Economists, value judgments, and climate change: a view from feminist economics. Ecol Econ 65(3):441–447
Nelson V, Meadows K, Cannon T, Morton J, Martin A (2002) Uncertain predictions, invisible impacts, and the need to mainstream gender in climate change adaptations. Gen Dev 10(2):51–59
Oreskes N, Conway EM (2008) Challenging knowledge: how climate science became a victim of the cold war. In: Proctor RN, Schiebinger L (eds) Agnatology: the making and unmaking of ignorance. Stanford University Press, Stanford
Scheman N (1980) Anger and the politics of naming. In: McConnell-Ginet S, Borker R, Furman N (eds) Women and language in literature and society. Praeger, New York
Seager J (2009) Death by degrees: taking a feminist hard look at the 2° climate policy. Kvinder, Køn, & Forskning 3:11–21
Slovic P (1999) Trust, emotion, sex, politics, and science: surveying the risk-assessment battlefield. Risk Anal 19(4):689–701
Smith JB, Schneider SH, Oppenheimer M, Yohe GW, Hare W, Mastrandrea MD, Patwardhan A et al (2009) Assessing dangerous climate change through an update of the intergovernmental panel on climate change (IPCC) "reasons for concern". Proc Natl Acad Sci USA 106(11): 4133–4137
Terry G (2009) No climate justice without gender justice: an overview of the issues. Gen Dev 17(1):5–18
Thomas CD, Cameron A, Green RE, Bakkenes M, Beaumont LJ, Collingham YC, Erasmus BFN, Ferreira de Siqueira M, Grainger A, Hannah L, Hughes L, Huntley B, van Jaarsveld AS, Midgley GF, Miles L, Ortega-Huerta MA, Peterson AT, Phillips OL, Williams SE (2004) Extinction risk from climate change. Nature 427(8):145–148
Traweek S (1988) Beamtimes and lifetimes: the world of high energy physicists. Harvard University Press, Harvard
Tschakert P, Tutu R, Alcaro A (2011) Embodied experiences of environmental and climatic changes in landscapes of everyday life in Ghana. Emot Space Soc. http://dx.doi.org/10.1016/j.emospa.2011.11.001
Tuana N (2006) The speculum of ignorance: the women's health movement and epistemologies of ignorance. Hypatia J Fem Philos 21(3):1–19
Tuana N (2012) Climate change and place: delimiting cosmopolitanism. J Specul Philos (in press)
United Nations (1992) United Nations framework convention on climate change. FCCC/INFORMAL/84. GE.05-62220 (E) 200705
Waring M (1990) If women counted: a new feminist economics. Harper Collins Publishers, San Francisco
World Bank (1989) Women in development: issues for economic and sector analysis working paper, population and human resources department i–95. http://go.worldbank.org/DR235FM3T0
Wylie A (2002) Thinking from things: essays in the philosophy of archaeology. University of California Press, Berkeley/Los Angeles

Chapter 3
A Climate for Feminist Intervention: Feminist Science Studies and Climate Change

Andrei L. Israel and Carolyn Sachs

Abstract For many years, climate change discourse was dominated by a technical-scientific framing based on modernist notions of objective knowledge, control, and efficiency. In recent years, a robust alternative discourse of climate justice has emerged, challenging mainstream adaptation and mitigation policies as reinforcing capitalist, colonialist, and patriarchal power structures and further marginalizing already vulnerable peoples and communities. But while the climate justice movement has provided a sorely needed corrective to climate change discourse, it has been hampered by addressing only policy issues without critically examining the scientific knowledge on which climate change discourse is based. Drawing on critiques of science and technology from ecofeminism and feminist science studies, we argue that scientific knowledge is always already structured by social power relations before it ever enters into policy discussions. In place of the (illusory) God-trick of absolute knowledge and control of the global climate system, we use Haraway's ideas of feminist objectivity, partial perspective, relations between species, and cyborg standpoints to situate and pluralize knowledge about climate change. This intervention opens up discursive space for multiple, partial knowledge about the climate system, all of which can be held accountable to their ethical and political implications. This pluralization of knowledge allows feminists to recognize and support many forms and venues of climate change-related activism, moving beyond the impasses of international and national political negotiations. Thus, far from dismissing climate change, a feminist critique of climate science makes possible a range of interventions that can more effectively promote social justice and ecological health.

A.L. Israel (✉)
Departments of Geography and Women's Studies, The Pennsylvania State University, University Park, PA, USA
e-mail: ali122@psu.edu

C. Sachs
Departments of Rural Sociology and Women's Studies, The Pennsylvania State University, University Park, PA, USA

M. Alston and K. Whittenbury (eds.), *Research, Action and Policy: Addressing the Gendered Impacts of Climate Change*, DOI 10.1007/978-94-007-5518-5_3,

Keywords Climate justice • Climate change • Discourse • Feminist analysis • Social justice

Feminist science studies offers a fresh approach to understanding and responding to climate change. Feminist responses to climate change have largely been framed within arguments about social justice, development, and global political economy, demanding attention to the inequities in impacts of and responsibilities for climate change (Denton 2002; Dankelman 2002; Lambrou and Pianna 2005; Terry 2009). These feminist voices argue that the impacts fall most harshly upon already-marginalized groups—such as people in the global South, subsistence farmers, poor women, and people of color—whereas the responsibility for greenhouse gas emissions lies overwhelmingly in the wealthy countries of the global North, especially the most powerful groups within those countries.

Given the urgency of these concerns, we argue that a feminist science studies analysis is essential to an effective feminist response to climate change. Certainly, we are not among the group of climate change deniers that reject or seek to cast doubt upon climate scientists' findings about the anthropocentric causes of global warming. As we will argue, rethinking (but certainly not *dismissing*) the climate science that grounds concerns about global climate change creates space for new ways of thinking about climate change and for new forms of activism, space that is urgently needed to formulate effective and just responses to the challenges of climate change.

We are motivated by the bleak state of climate politics. Within the U.S., levels of public skepticism about climate change are high and continue to increase. On the international level, negotiations on greenhouse gas emissions reductions move forward slowly if at all. We argue therefore that radically new ways of thinking, talking, and acting about climate change are needed if we are to respond effectively, not to mention sustainably and justly. To explore such possibilities, we must take the risk of politicizing the science of climate change, breaking down the divide that separates climate science from climate politics.

In this paper, we propose a broad agenda for feminist intervention into climate change discourse. Feminist scholars of gender and development have provided cogent analyses of the gendered impacts of climate change, arguing (in line with the broader "climate justice" movement) that climate change is not only an environmental issue but also an urgent social justice issue. But this "gender and climate change" literature intervenes only in social responses to anthropogenic climate change, without addressing the processes by which scientific knowledge about climate change is produced and distributed. Drawing on the theoretical resources of feminist science studies, we argue that the scientific knowledge about climate change can and must be critically evaluated and reconstructed. Indeed, the centrality of mathematical and technological science to climate change discourse in part explains the scarcity of feminist contributions to this discourse, given the extent to which these forms of inquiry are gendered as masculine. To bracket off "the science" from critique removes from scrutiny the ways in which climate science is structured by masculinist ideologies of domination and mastery, as well as the ways social inequalities determine who does and does not participate in producing scientific knowledge about climate change. In the tradition of feminist environmentalism, we call for new ways of knowing about and responding to climate change that take seriously the materiality of the phenomenon but also interrogate how climate change is inextricably wrapped up in social relations of power and oppression. We borrow from Donna Haraway's analyses of science to articulate a view of climate science as situated knowledge, always partial and limited (and never innocent) but also valuable and useable in its partiality. The ethical and political issue therefore becomes not only how to promote social justice in responses to climate change, but how to act (ethically, and in solidarity with those suffering from the effects of climate change) given the uncertainties inherent in social and environmental systems and the impossibility and undesirability of total knowledge or control.

We conclude by suggesting three directions for new feminist programs of research and activism. The first direction is deconstructive and oppositional: feminists must oppose and intervene against the logics of domination and control that are far too common in climate change discourse and policy, supporting those marginalized voices of dissent against, for example, ill-conceived geo-engineering proposals or attempts at large-scale social control. The second approach we recommend involves closer feminist analysis of climate science itself. To support effective and just action on climate change, we need a much more robust understanding of the social and technical processes that produce knowledge about the global climate system. Such analyses will open spaces for activism on the politics of knowledge, supporting feminist efforts to re-value the local environmental knowledges that have been thoroughly marginalized in climate change discourse. Finally, we propose that a feminist response to climate change requires new approaches to environmental ethics and activism that go beyond both the managerial utilitarianism of mainstream environmental activism and a purely critical stance that can impede meaningful action.

Climate Change Science and the Construction of Environmental Concern

The origin and emergence of concerns about climate change have followed a very different course from many other environmental issues. Many environmental issues have gained prominence through the efforts of heterodox scientists (e.g., Carson 1962) or lay (non-scientist) activists who challenged the scientific and political mainstream, raising concerns about activities that were thought to be harmless, such as in the Love Canal incident (Newman 2001). In contrast, concerns about climate change first came directly from the scientific establishment (Stehr 2001). The physics of the greenhouse effect—namely, that certain gases in the atmosphere (most importantly, water vapor, carbon dioxide, and methane) trap heat and thereby keep the planet much warmer than it would be in their absence—have been understood since the 19th century, and the idea that human activities such as deforestation and combustion of fossil fuels might impact the climate first emerged soon after (Edwards 2010). But it was only with the post-World War Two development of numerical modeling techniques and digital computers that scientists were able to evaluate the effects of increasing atmospheric concentrations of carbon dioxide on climate and raise concerns about global warming.

Because of these beginnings, discussions of climate change have been structured from the start by a techno-scientific framing (Demeritt 2001). Climate change science has focused its energy on the construction of ever-more-precise accounts of the dynamics of the global climate system and the ecological and social systems that are intertwined with it, in efforts both to predict future climate conditions and to use that understanding to control those conditions. Indeed, modern climate science grew directly out of initiatives that sought to automate (and thereby improve the accuracy of) weather forecasting and to manipulate weather conditions through, for instance, "seeding" clouds to increase rainfall (Edwards 2010). The sorts of social questions that were common in the context of other environmental issues—concerning, for example, the social conditions that produce greenhouse gas emissions, the effects of changing climate conditions, or the limits to science's ability to provide technical solutions to the problem—were not raised during the early emergence of climate change concerns, not surprisingly since it was scientists and engineers, not activists, who raised the concerns.

This techno-scientific approach to thinking about climate change is seen clearly in the work of the Intergovernmental Panel on Climate Change (IPCC). The IPCC is divided into three working groups, addressing, respectively: (1) the physical science basis of climate change; (2) impacts, adaptation, and vulnerability; and (3) mitigation of climate change. The prominent positioning of physical-scientific knowledge in Working Group one (WG I) is no accident: the physical-scientific accounts of climate change in the WG I report are the basis of the IPCC report as a whole. That is, discussions of impacts and mitigation in the reports of Working Groups two and three (WG II and WG III) are based on the analyses in the WG I report.

The report of WG I "provides a comprehensive assessment of the physical science of climate change" (IPCC 2007a, vii). The research synthesized in this report falls into three basic categories: empirical observations of the composition of the atmosphere and the state of the climate in the present and recent past; theoretical accounts (often supported by paleoclimatic data) of how the climate system responds to various drivers, including (but not limited to) varying concentrations of carbon dioxide and other greenhouse gases; and projections of future climates from mathematical models of the global climate system (General Circulation Models, or GCMs). But this apparent balance among observational, theoretical, and predictive aspects belies the organizing purpose of the report: to provide "the most complete and quantitative assessment of *how human activities are affecting* the radiative energy balance in the atmosphere" (IPCC 2007a, i, emphasis added). In short, the goal of the WG I report is to determine how anthropogenic emissions of greenhouse gases are affecting the climate system. Mathematical climate models, with their ability to project future climate under various emissions scenarios and to construct a hypothetical present without anthropogenic emissions, are the essential tool for achieving this goal (Demeritt 2001). Thus, the observational and theoretical components serve primarily to provide inputs and ground-truth checks for the GCMs; it is only via the GCMs that observed changes in climatic conditions can be attributed with confidence to anthropogenic influences.

For the IPCC's purposes, the existence and nature of anthropogenic climate change is established in the WG I report. The reports of Working Groups II and III, then, address the "so what?" questions of why climate change might matter (to people and ecosystems) and how we might respond to it. The WG II report (IPCC 2007b) addresses the possible consequences of this climate change for natural and human systems and the prospects for adaptation of these systems to altered climate conditions; the WG III report (IPCC 2007c) addresses strategies for mitigating future climate change through stabilization of atmospheric concentrations of greenhouse gases. In both reports, the focus is to articulate the physical science from the WG I report with research on the structure and functioning of ecological and social systems.

To summarize, the IPCC's techno-scientific approach to climate change presents a linear path through the following steps: first, determine the existence and nature of anthropogenic changes in the global climate system (WG I); second, determine the impacts of these changes on ecological and social systems (WG II) and the social sources of these changes (viz., greenhouse gas emissions—WG III); third, determine how these systems might adapt to change in climate (WG III) and how human activities that cause greenhouse gas emissions might be altered; fourth (and finally), implement policies to support both adaptation efforts and emissions reductions. The IPCC report itself stops at step three, with policy implementation left for political frameworks such as the United Nations Framework Convention on Climate Change (UNFCCC) and related processes such as the Conference of Parties (CoP) meetings.

The Consequences and Limitations of a Techno-Scientific Approach

Although not in the context of climate change, feminist scholars have explored thoroughly the effects of a techno-scientific approach to social and environmental issues, and the experience with climate change bears out much of this analysis. In climate change discourse, the physical status of the global climate system is treated as a straightforward fact of nature, separate from the political, cultural, and ethical questions of what society can or should do about it. Within this frame, questions about whether or how the climate is changing can be addressed only through the established procedures of the scientific community—that is, within the "republic of science" (Polanyi 1962) rather than in wider society.

Feminist epistemologist Lorraine Code argues that such an approach relies upon an "implicit claim... that empirical inquiry is not only a neutral and impersonal process, but also an inexorable one; it is compelling, even coercive, in what it turns up to the extent that a rational inquirer *cannot* withhold assent" (Code 1993, 28, emphasis original). Code's account destabilizes this view of epistemology, arguing against the "impersonal" view of scientific inquiry to show how the identities and social positions of researchers shape their claims to knowledge. Further, she argues that the scientific approach to knowledge implicitly but effectively excludes women and people from other socially-marginalized groups from being recognized as legitimate knowers (see also Code 1991). More importantly, though, her argument shows how a scientific epistemology[1] insists on the separation of facts from values and politics. Feminist science studies scholars have argued that this feature of scientific epistemology enables the concealment of social power relations within purportedly apolitical scientific knowledge claims. For instance, the patriarchal association of femaleness and passivity can be seen in 18th century botany (Schiebinger 1993) and 20th century reproductive biology (Martin 1991). These sorts of insights cast Code's point about the coercive nature of scientific epistemology in a different light: in practice, this coercion originates not in the brute facticity of the evidence, but rather in the power-laden social context that supports patriarchal beliefs.

These feminist critiques of science seem less relevant to climate change science and its current epistemological politics. In the above examples (and countless others explored by feminist scholars), hegemonic and conservative social forces operate by enforcing belief in (supposed) natural facts of, for example, women's inferior mental and physical abilities. With climate science, in contrast, conservative social forces are aligned *against* the scientific consensus, promoting not acceptance but skepticism of the scientific facts in efforts to secure the hegemonic social order (McCright and Dunlap 2010; Oreskes and Conway 2010). One common reaction in this situation would be to "flip the script" on the critique of science, staunchly defending the claims of climate scientists and rejecting the skepticism that underlies the sorts of feminist analyses described above.[2] But if we take a wider view of the epistemological politics of climate change, the situation actually exemplifies the problematic nature of modern science that the feminist critiques have illustrated.

[1]Some feminist and post-colonial scholars would insist on speaking not of science in general, but of modern Western science in particular, as opposed to non-Western or indigenous knowledge practices (see, e.g., Harding 1991, 2008). We agree that this is an important distinction, but for the sake of clarity and simplicity we use the term "science" to describe the modern, Western approach to knowing about the natural world.

[2]See, for example, Bruno Latour's oft-cited lament about the politics of critique in the introduction to his essay "Why has critique run out of steam" (Latour 2004).

That is, we argue that rather than an entrenched defense of science against its conservative adversaries, what is needed is a re-thinking of scientific practice itself.

The present situation regarding climate change clearly illustrates two consequences of the scientific epistemology that Code and other feminist scholars have described, and particularly of its insistence on a clean separation between facts and values. First, while the purpose of this separation is to keep science safe from political manipulation, in practice, "the walls [defending science] are turning out to be made of straw" (Collins 1983, 100). That is, a *political* debate is being waged on epistemological terrain when conservatives challenge the legitimacy of the scientific consensus on climate change, but this is discursive terrain on which arguments about justice and ethics have little room to stand. Going with the feminist insight that epistemology is political through-and-through, it is clear that simply challenging climate change skeptics on their facts misses the point and holds little chance of succeeding: empirical evidence is never adequate to close a scientific controversy on its own, and given the political stakes regarding climate change, we can and should expect continued efforts to cast doubt about the climate change science.

Beyond the politically-charged epistemological impasse, the separation of scientific facts from values and politics has a second, equally dire, consequence in the context of climate change. Because climate change discourse is focused on abstract, technical issues such as modeling practices, atmospheric physics, and cumulative greenhouse gas emissions, climate change is detached from everyday life experiences, making it more difficult to mobilize political will around climate change (Jasanoff 2010). That is, rather than a straightforward translation of scientific facts into political debates and decisions, scientific knowledge on climate change has had difficulty generating political urgency and action, in the face of more clearly political issues such as the world-wide economic crisis hold sway.

In practical terms, the consequence of the scientific framing of climate change is that the dominant mitigation and adaptation policies generally focus on technical issues that can be managed through environmental or social engineering. In these measures, the problem of climate change is narrowly understood as only involving the Earth's energy balance[3] or the composition of the atmosphere, leading to solutions that neglect the social causes and complexities of the issue. One extreme example of a technical solution is the investigation of "geo-engineering" techniques like placing giant mirrors in space to deflect Earth-bound solar radiation, thereby (in theory) counteracting the warming of the atmosphere (Schneider 1996). But this approach is equally seen in schemes to reduce carbon dioxide emissions through taxation or commoditization, as in the Kyoto protocol's system of tradable "emissions credits" (Zhang 1998). In both cases, the response to climate change seeks to control social and environmental systems to avoid or lessen the risks of climate change; fundamental change to society such as halting the fossil fuel economy, reducing consumption, and addressing inequities in contributions to global warming are barely on the table.

[3]The "energy balance" is the difference between solar energy coming into the atmosphere and energy radiating from Earth out into space. If the energy balance is positive, the atmosphere is trapping more heat and therefore becoming warmer.

Given these consequences of the technical/scientific approach to climate change, we argue—*contra* the common calls for a *de*-politicization of climate change science (e.g., Giddens 2009)—that a feminist response to climate change must *re*-politicize the science, emphasizing the interconnections between scientific knowledge and practice and issues of social justice and social change. Such a new approach to climate change, we argue, is crucial to supporting effective and socially just responses to climate change. In the next section, we review the growing literature on climate justice and gender and climate change, which represents a promising start to our project, foregrounding precisely those dimensions which are excluded by the narrow technical approach to adaptation and mitigation—namely, unequal social power relations and a need for social justice. We then turn our attention to climate change science itself, using the resources of feminist science studies to suggest ways of reconstructing climate change science along feminist lines. We conclude by suggesting directions for feminist research and activism that could promote both productive reframing of climate change and meaningful action to reduce the vast social and ecological harms being caused by climate change.

Environmental Justice, Climate Justice, and Gender Justice

In recent years, a counter-weight to mainstream technical approaches to climate change has emerged, in the form of the climate justice movement (Shepard and Corbin-Mark 2009). This emerging movement has its roots in the environmental justice movement. In the U.S., the environmental justice movement began in the early 1980s as predominantly African-American communities recognized that African-American neighborhoods and communities were more likely to be exposed to toxic substances and wastes due to siting of waste facilities, landfills, and toxic chemical plants in or near their communities (Cutter 1995). The environmental justice framework insists that race, ethnicity, and poverty shape inequalities in the impact of environmental degradation. Recently the environmental justice movement has focused on the implications of climate change for vulnerable people and communities and called for attention to climate justice. For instance, poor communities are most likely to be affected by the natural disasters and extreme heat events that climate change makes more likely (Shepard and Corbin-Mark 2009). In the U.S., the most impacted communities will be communities with high proportions of people of color that are already socially, economically, and environmentally disadvantaged. In a particularly vivid example, the experiences of residents of New Orleans and the broader Gulf Coast with Hurricane Katrina clearly exemplified the increased vulnerability of poor African Americans to storms. Climate justice advocates point out that the world's poorest people will be most vulnerable to climate change due to their location in areas such as coastal zones, semi-arid lands, flood plains, and other especially vulnerable locations, as well as their lack of economic resources to adapt to climate change. While climate change is a global phenomenon, certain nations, particularly in the global South, are currently experiencing and are more likely to suffer from global warming. Many island nations in the Pacific and many African nations are most vulnerable to rising global temperatures. Climate change negotiations have been stymied by disagreements between Northern and Southern countries partially over the question of the ecological debt that the North owes the South (Parks and Roberts 2010). Many Southern countries and NGOs argue that the countries of the global North must make concessions and repayment of this ecological debt due to their role in decades of exploitation of the natural resources and ecosystems of poor countries (Martinez-Alier 2003; Sims et al. 2003). Justice and fairness issues are often negotiated at the global level between states such as the IPCC meetings. But agreements of justice and fairness are difficult to achieve because, as Parks and Roberts (2010, 148) suggest, we live in "morally ambiguous" times with notions of fairness in climate change varying by circumstance with some countries blaming their situation on Northern consumption, others desiring to develop their economies using a carbon-based economy, and still others such as small island nations calling for an end to carbon trading. They suggest that inequality between the North and South makes agreement on a fairness principle unlikely and that consensus will require "a hybrid justice solution that accommodates the different circumstances and principled beliefs of many parties" (2010, 152).

The emerging climate justice framework tends to focus at the nation-state level and on indigenous peoples in vulnerable locations. More recently feminist interventions have called for the inclusion of gender in the climate justice debates. The initial concern with gender and climate change emerged from scholars and policy makers working on gender, environment, and development (Dankelman 2002; Denton 2002). They argue that gender matters for climate change. These feminist scholars see gender as the primary axis that structures difference and emphasize how gender intersects with other statuses and identities. These insights suggest that the intersection of gender, race, ethnicity, and economic status differentially position people in terms of their relationship with the environment. Access to, use of, and control over land, water, forests, and other natural resources vary by gender in most contexts (Dankelman 2002). With the degradation of environmental resources such as water and forests, poor women are often most disadvantaged because their daily activities rely heavily on accessing food, water, and energy (Denton 2002). Because women more frequently live in poverty than men, they are more vulnerable to both large-scale disasters such as floods as well as slow-onset impacts of climate change such as drought and rising temperatures. Poor women's and girl's daily activities frequently involve accessing water and fuelwood; degradation of these resources through climate change often increases their workload. Also, in times of disaster and relocations, women and girls are increasingly susceptible to sexual and domestic violence.

Many of the gender and climate change scholars are careful to insist that women are not only victims of climate change, but also often cope and adapt to disasters differently than men using their particular knowledge and livelihood strategies, for instance by changing the crops they grow or switching to subsistence rather than commodity production. Gender and development scholars and policy makers and other advocates of social justice argue that climate is only one of many factors impacting women's and poor people's lives (Lambrou and Pianna 2005).

These interventions to foreground issues of social justice and gender justice in climate change discourse represent an important first step in a feminist reframing of climate change, creating needed space for strategies that simultaneously address issues of development and climate change. But this movement only politicizes the *impacts* of climate change: the scientific understanding of climate change remains bracketed, forming the basis for climate justice activism but not itself subject to critical analysis. To create ways out of the current impasse, climate change science must be politicized, and it is to this project that we now turn.

Politicizing Climate Science: Rethinking Climate Change in Feminist Terms

In our efforts to rethink climate science in feminist terms, we join a growing number of feminist science studies scholars calling for a shift from debunking or resisting scientific claims toward a more interactive engagement with science and scientists, taking seriously the knowledge about the natural world that science can provide while examining the political and social dimensions of scientific knowledge and practice (see, e.g., Kirby 2008; Wilson 2002; Grosz 2008). Donna Haraway's (1988) critical analysis of technoscience serves as a foundation for this reconstructive project. Haraway raises sharp concerns about the ideological underpinnings of the scientific project of seeking absolute, universal knowledge, arguing that this project is thoroughly entangled with militaristic, masculinist, and colonialist ideologies. She describes the scientific ideal of absolute, universal knowledge as a "God-trick of seeing everything from nowhere," whose effect is "to distance the knowing subject from everyone and everything in the interest of unfettered power" (1988, 581). The IPCC's mission of "providing policymakers with objective scientific and technical findings that are policy-relevant but not policy prescriptive" (2007c, v) is a practical manifestation of this approach to scientific knowledge, as the knowers (scientists) are construed as removed from the messy political and ethical terrain in which policymakers (and others) must act. As we have argued above, this framing of climate science does advance the cause of "unfettered power"—at least insofar as it leads to technical "solutions" that involve increasing control of social and environmental systems. The IPCC's multi-colored maps of future climate conditions, which have become common representations of climate change, are a vivid example of this "god-trick," representing not only present but future climates as thoroughly knowable and therefore (implicitly) controllable.

But the power of Haraway's analysis is that she goes beyond the self-representation of science and scientists to reexamine and reframe scientific practices themselves. This examination is based on two concepts: feminist objectivity and cyborg embodiment.[4] Haraway defines feminist objectivity as follows: "feminist objectivity is about limited location and situated knowledge, not about transcendence and splitting of subject and object" (1988, 583). Unlike the universal knowledge that modern science purports to provide (which in fact just represents the imposition of power from dominant social positions), feminist objectivity recognizes that all knowledges reflect a partial perspective and acknowledges and even embraces its partiality. Against its own claims, then, Haraway would see the scientific knowledge about the climate reported by the IPCC as situated knowledges, unavoidably partial (both in the sense of "incomplete" and in the sense of "biased").

Cyborg embodiment comes into play when Haraway pushes her feminist version of objectivity beyond mere recognition of partiality. She argues that feminist objectivity is achieved not through separation from objects of knowledge but through embodied connection and engagement with them: "objectivity turns out to be about particular and specific embodiment" (1988, 582). The cyborg, Haraway's (1985) ironic figure for feminist politics, provides a way to understand this "particular and specific embodiment" in the context of climate change science. The cyborg is a hybrid being, part human, part cybernetic machine, and therefore always constituted through the processes of capitalist / masculinist technoscience. But Haraway does not therefore see the cyborg as compromised and useless to feminist activism. Instead, she celebrates cyborgs as representing a non-innocent position within which one might claim knowledge and take action. Thus, in her essay on "Situated Knowledges" (1988), when she elaborates that *all* perceiving and knowing is technologically mediated, she is invoking a cyborg embodiment. That is, rather than separating direct, embodied experience from technologically-mediated ways of knowing, Haraway conflates them, arguing that the body itself is a semiotic-material technology of knowing and acting: "embodiment is significant prosthesis" (1988, 588).

Following Haraway's lead, a feminist reconstruction of climate science must see climate models and their projection maps as one form (among many) of technologically mediated, embodied, and situated knowledge about the climate. Karen Barad's (2007) concept of an "apparatus" helps make sense of the process: in Barad's terms, the models, supercomputers, satellites collecting data, etc. all constitute an apparatus for producing knowledge about the climate. Further, the scientists themselves, institutions like the IPCC, and a whole range of other social entities are integral parts of this apparatus. But the key point of Barad's framework is that an apparatus is not only a means of knowing about a pre-existing phenomenon; rather, the apparatus constitutes the phenomenon itself, and does so in specific ways. The "global climate" only exists as an object of concern for us today because of the complex socio-technical apparatuses of climate science; thus, it is not a question of finding ways to know about the climate that are free of the hegemonic

[4]The phrase "cyborg embodiment" is our own, not Haraway's, but we think captures well the way the figure of the cyborg might matter to scientific knowledge, especially in as technological a field as climate science.

tendencies of modern science, but rather of recognizing the partiality and limits of any representation of the climate—both in terms of what is left out or glossed over and in terms of what the representation performatively constructs—and of assessing the sorts of ethical and political response that any particular version of "global climate" enables or obstructs.

One crucial and problematic aspect of mainstream climate science is the issue of scale. Climate change science speaks primarily about global-scale processes and effects, thereby discursively producing an "undifferentiated . . . global citizenry" as the victim of climate change (Demeritt 2001, 313). Thus, this dominant framing of the issue renders invisible the ways in which different communities and places experience climate change differently and specifically how (as discussed above in relation to the climate justice argument) poor and marginalized people are especially vulnerable. Moreover, the discourse of an undifferentiated global citizenry obscures the wildly different levels of responsibility for greenhouse gas emissions between the global North and South, thereby writing out of climate ethics and politics any reference to differences in historical emissions, wealth, and political-economic power. In place of the totalizing, global view of mainstream climate science, then, feminist interventions into climate change science must make visible the differentiated local effects and causes of climate change, producing knowledges that support ethical and political action. Further, this global-scale framing actually impedes public understanding of and response to climate change: "Appeal to the universal interests of a global citizenry is founded on scientific certainty, rather than the more difficult work of making global warming meaningful to a differentiated international public" (Demeritt 2001, 329). In fact, several climate change scientists recently discussed the limits of climate science models in predicting local climate futures or providing communities with projections or appropriate scale of information for adapting to climate change (Crane 2011). Accordingly, in place of the univocity of technical/scientific frame, multiple ways of representing and relating to climate change are needed if the issue is to be meaningful, and therefore matter to ethical and political decisions for diverse people in a variety of situations around the world.

For a suggestion of a feminist approach to thinking and acting in the face of the uncertainties of climate change, Haraway's recent work on human-animal relationships provides a starting place. In *When Species Meet* (Haraway 2008), Haraway addresses the question of how to act ethically in the context of partial and situated knowledges. Given the uncertainties of climate change, the varied local implications, and the long time horizon involved, partial knowledges necessarily abound. Rather than being paralyzed by the partiality of knowledges, Haraway calls for the cultivation of "response-ability", the ability and willingness to respond to what we learn. In particular, she calls for habits of curiosity and compassion—responding to new knowledge by seeking to learn more, by feeling for those about whom you have learned, and by acting, insofar as possible, to reduce suffering.

Response-ability serves as an alternative to and intervention against the cut-and-dried strictures provided by rationalist-instrumentalist ethical frameworks. Response-able action involves an empathetic attunement to the needs of an other, including non-human others, in the face of uncertainty about what the ethically "right" action is. In these terms, the attitude of scientific detachment produced by technocratic discourses on climate change actually impedes response-ability because it prevents affective connection with the causes and consequences of changing climates. Thus, an ethics of response-ability goes beyond the scientific framework that both the IPCC and the climate justice movement work within, instead cultivating a sense of connection and interdependency and a capacity to act ethically in the context of (inevitable) uncertainty.

New Climates: Feminist Research and Activism on Climate Change

We have argued that the dominant scientific and technical framing of climate change discourse and policy prevents effective and just action in the face of global change. Opening up discussion and investigation of the science that lies at the foundation of climate change discourse may seem like a dangerous move, one that could support the political forces of climate change denial, which seek to continue the business-as-usual capitalist exploitation of the working class, women, and the natural environment. But feminist theory has a long history of challenging the taken-for-granted, as in Butler's (1990) challenge to given gender/sexuality categories or Gibson-Graham's (1996) questioning of the discourse of capitalist hegemony. By all accounts, current efforts toward climate change mitigation and adaptation have little hope of responding adequately to the problem—very little serious progress has been made toward reducing greenhouse gas emissions through the Congress of Parties negotiations and other orthodox political processes. In this context, we argue that feminists must take this risk of further opening up the discussion, and we suggest that this opening up may create myriad new paths and sites for feminist action. We suggest a program of feminist research and activism on climate change that extends well beyond the important work already being conducted within the "gender and development" and "gender and disaster" frameworks (Dankelman 2002; Enarson and Meyreles 2004). A wide range of theoretical, empirical, and activist projects are needed, in a variety of sites and contexts. In this final section, we outline three key strands of feminist intervention in the realm of climate change.

Challenging discourses of control and domination in climate science and policy – Feminists must challenge and resist the techno-scientific framing of climate change, which relies on an ideology of the control and domination of nature. A healthy and just world cannot come out of hubristic efforts to control what is ultimately uncontrollable. Joni Seager's (2009) critique of the 2°C "global warming target" illustrates the power of this feminist critique, as she argues compellingly that the idea that global warming can be stopped at a precise level is both a fantasy of domination and cover for protecting the interests of the powerful. The point, Seager argues, is not to carefully manage anthropogenic damage to the natural environment, but rather to *prevent* such damage.

As a specific practical example, feminist theoretical frameworks in science studies and environmental studies provide solid grounding from which to resist the growing interest in geo-engineering approaches to mitigating climate change. Where geo-engineering reduces climate change to a matter of energy balance in the atmosphere—solvable by reflecting more energy back into space (e.g., with mirrors in space) or by removing greenhouse gases from the atmosphere ("carbon capture and storage" underground)—feminists must insist upon the importance of the complexity and uncertainty concealed within that reduction. Atmospheric scientists themselves have pointed out several problems with geo-engineering approaches, including uncertainty regarding the effectiveness of various interventions, the need for ongoing (and indeterminate) funding and political support, unevenness of global impacts, and the likelihood of unforeseen negative consequences (Robock 2008). To these analyses, feminists can contribute a rigorous account of the limitations of *any* project to control natural and social systems, as well as raising awareness of the likely uneven distribution of burdens when these projects go awry.

Given the impossibility of God-like foreknowledge of or control of climate futures, feminist approaches to climate change must accept uncertainty and unpredictability. Interventions must be as open-ended as possible and must be constantly re-evaluated critically as circumstances change. Such an approach will look radically different than the technical/scientific versions of environmental planning and management that have thus far failed effectively to address climate change.

Exploring the partialities of climate change science – Haraway's concept of "situated knowledges" provides a way for feminists to understand the limitations of climate science without rejecting it altogether. Empirical studies can contribute valuable insights into specifically *how* knowledges about the global climate are situated and partial. Research that follows the actors (Callon et al. 1986) through the production, circulation, and application of climate science can provide important insights into how climate change science works, revealing (for instance) who participates in these processes, where and how they occur, and how the knowledge produced and policy decisions taken are impacted by the inclusions and exclusions of these processes. For example, Bowser (2010) argues that the needs and perspectives of people of color have been largely absent from climate science, in part because people of color have been relatively absent from the scientific research and policy processes. Historical or ethnographic studies, modeled on Latour and Woolgar's (1979) study of biomedical research or Traweek's (1988) study of theoretical physics, of the technologies involved in climate science (supercomputing, satellite imagery, modeling, etc.) could illuminate the assumptions built into these tools and the partialities of what such tools can tell us.

This kind of critical analysis—exposing the assumptions and limitations of climate science—will create space for new knowledges about climate, knowledges that come out of different social (and geographical) locations and from different tools than mainstream climate science. As Harding (2008) reminds us, people typically excluded from Western science—women, people of color, and people of the global South—have their own ways of knowing about the natural world. Feminists must insist upon the value of these multiple knowledges for effective action concerning climate change. Petra Tschakert's use of participatory research to assess and build adaptive capacity with agriculturalists in sub-Saharan Africa provides a promising example of such an approach (Tschakert 2007; Tschakert and Dietrich 2010). In this work, the technoscientific knowledge of Western climate science plays a role, but the rich knowledge of local climate history that rural agricultural people already possess is equally valued as the foundation for community adaptation to changing climates. In this and related ways, feminist climate change activism must reject Western science's insistence on "one true story" (Stanley and Brickhouse 2001, 46), instead recognizing the multiplicity of perspectives on climate as well as both the utility and limitations of each perspective for responses to climate change.

Facilitating response-able action – These twin epistemic moves—decentering the authority of scientific accounts of the climate system (by situating them within power-laden social, geographic, and historical contexts) and affirming multiple knowledges about climate—must not open the door for self-interested climate change skeptics and their rich and powerful backers. The payoff for these moves comes in new possibilities for response-able feminist action on climate change.

What sorts of interventions could promote response-able action on climate change? Clearly there can be no one "magic bullet" response—pluralism and open-endedness, rather than closure and univocity, are key. Any process must therefore be as inclusive and participatory as possible. Space must be made for creative, experimental responses—perhaps responses that include a sense of playfulness, rather than the dour earnestness that typically characterizes climate change negotiations. For example, Alaimo (2009) describes a recent protest involving hundreds of people standing naked on a melting iceberg. Such protests build on past feminist actions to resist militarization and environmental destruction, such as the Women's Pentagon Action in Washington, D.C. and the Greenham Commons Women's Peace Camp in Berkshire, England, which mobilized numerous women to resist possible annihilation in the face of nuclear military technology. But the recent naked protests represent a new twist on this activist tradition, viscerally representing vulnerability to climate change and to ecological systems more generally through nakedness, by "perform[ing] vulnerability as a trans-corporeal condition, in which the material interchanges between human bodies, geographical places, and vast networks of power provoke ethical and political actions" (Alaimo 2009, 23). Seeing this kind of vulnerability as part of an inescapable connection to the more-than-human, rather than a terrifying prospect to be avoided, creates space for affective response and perhaps then for responsive action.

Within the framework of pluralistic and experimental responses to climate change, a crucial feminist contribution must be an insistence on holistic solutions that promote both environmental *and* social well-being. Feminist environmentalism has long held that social injustice and ecological devastation are two sides of the same coin (e.g., Shiva 1988; Seager 2003); feminist responses to climate change must likewise resist approaches that neglect social justice issues as both ineffective and ethically unacceptable. Further, a necessary part of such holistic approaches will be expanding the field of "climate change responses" to include activities at many different scales and sites. The present framing of climate change action as necessarily global and limited to orthodox political processes (i.e., international negotiations and national legislative processes) has produced an apparent contradiction between economic justice and ecological health (Parks and Roberts 2010); in this context, holistically-minded interventions may be essential at local or regional scales, or through social processes outside of orthodox politics. Naked protests are one example of alternative sites of climate action; another example is Interfaith Power and Light, a growing U.S.-based movement of religious congregations to conserve energy and promote renewable energy sources (http://interfaithpowerandlight.org/). Tackling climate change at an explicitly local scale, the Transition Town movement (http://www.transitiontowns.org/) seeks to "relocalize" life in cities and towns, aiming to reduce use of fossil fuels, mitigate greenhouse gas emissions and other harmful environmental impacts, and build stronger local economic and social networks.

These few examples show that in spite of the dominance of techno-scientific discourses that frame climate change in reductive terms and exclude non-conformant perspectives, instances do exist of climate change action outside this narrow frame. Feminist scholars and activists must play a key role in these movements, creating discursive space for the alternatives they represent, helping to navigate the fine lines between creative protest and attention-seeking spectacle (Alaimo 2009) or between politics done differently and a retreat from politics. Moreover, the participation of feminists in these movements is crucial if they are to truly implement anti-hegemonic and inclusive activities, given feminist sensibilities about the subtle workings of power and privilege. While women are often taking the lead on climate change action at local levels, feminists must also engage in global deliberations and policy. Local action will not suffice. Feminists must insist that gender issues be on the table of global policy making such as the IPCC and the Conference of Parties. New strategies and new forms of feminist organizations are needed to incorporate local, situated knowledge while simultaneously engaging with broader scientific agendas and global policies and practices.

Feminist scholars and activists are increasingly recognizing environmental issues broadly, and climate change specifically, as central social justice issues of the twenty-first century. As feminists turn greater attention toward these issues, we must not simply accept "the science" as the apolitical ground of political action. As we have argued in this paper, the traditions of feminist science studies and environmental studies can provide essential theoretical resources for understanding climate change science, politics, and policy as complex socio-environmental phenomena, involving unequal power relations among a wide range of actors. Given the failure of mainstream efforts to respond effectively to climate change, these transformative insights and interventions are urgently needed.

References

Alaimo S (2009) Insurgent vulnerability and the carbon footprint of gender. Kvinder, Køn & Forskning 3–4:22–33

Barad K (2007) Meeting the universe halfway: quantum physics and the entanglement of matter and meaning. Duke University Press, Durham

Bowser G (2010) Climate and equity: ecological implications of climate change on communities of color. Presentation at annual meeting of the Ecological Society of America, Pittsburgh, 3 August

Butler J (1990) Gender trouble, feminist theory, and psychoanalytic discourse. In: Nicholson LJ (ed) Feminism/postmodernism. Routledge, New York, pp 324–340

Callon M, Law J, Rip A (eds) (1986) Mapping the dynamics of science and technology: sociology of science in the real world. Macmillan, Basingstoke

Carson R (1962) Silent spring. Houghton Mifflin, Boston

Code L (1991) What can she know? Feminist theory and the construction of knowledge. Cornell University Press, Ithaca

Code L (1993) Taking subjectivity into account. In: Alcoff L, Potter E (eds) Feminist epistemologies. Routledge, New York, pp 15–48

Collins HM (1983) An empirical relativist programme in the sociology of scientific knowledge. In: Knorr-Cetina KD, Mulkay M (eds) Science observed: perspectives on the social study of science. Sage Publications, London, pp 85–113

Crane R (2011) Crash course on climate down-scaling. Presentation at limits to climate change adaptation workshop, University Park, 8 June

Cutter SL (1995) Race, class and environmental justice. Prog Hum Geogr 19(1):111–122. doi:10.1177/030913259501900111

Dankelman I (2002) Climate change: learning from gender analysis and women's experiences of organising for sustainable development. Gend Dev 10(2):21–29

Demeritt D (2001) The construction of global warming and the politics of science. Ann Assoc Am Geogr 91(2):307–337. doi:10.1111/0004-5608.00245

Denton F (2002) Climate change vulnerability, impacts, and adaptation: why does gender matter? Gend Dev 10(2):10–20

Edwards P (2010) A vast machine: computer models, climate data, and the politics of global warming. MIT Press, Cambridge, MA

Enarson E, Meyreles L (2004) International perspectives on gender and disaster: differences and possibilities. Int J Sociol Soc Policy 24(10/11):49–93. doi:10.1108/01443330410791064

Gibson-Graham JK (1996) The end of capitalism (as we knew it): a feminist critique of political economy. Blackwell Publishers, Oxford

Giddens A (2009) The politics of climate change. Polity, Cambridge

Grosz E (2008) Chaos, territory, art: deleuze and the framing of the earth. Columbia University Press, New York

Haraway DJ (1985) Manifesto for cyborgs: science, technology, and socialist feminism in the 1980s. Social Rev 80:65–108

Haraway DJ (1988) Situated knowledges: the science question in feminism and the privilege of partial perspective. Fem Stud 14(3):575–599

Haraway DJ (2008) When species meet. University of Minnesota Press, Minneapolis

Harding S (1991) Whose science? Whose knowledge? Thinking from women's lives. Cornell University Press, Ithaca

Harding S (2008) Sciences from below: feminisms, postcolonialities, and modernities. Duke University Press, Durham

IPCC (2007a) In: Solomon S, Qin D, Manning M, Chen Z, Marquis M, Averyt K, Tignor MMB, Miller HL (eds) Climate change 2007: the physical science basis. Contribution of working group I to the fourth assessment report of the intergovernmental panel on climate change. Cambridge University Press, Cambridge

IPCC (2007b) In: Parry M, Canziani O, Palutikof J, van der Linden P, Hanson C (eds) Climate change 2007: impacts, adaptation and vulnerability. Contribution of working group II to the fourth assessment report of the intergovernmental panel on climate change. Cambridge University Press, Cambridge
IPCC (2007c) In: Metz B, Davidson O, Bosch P, Dave R, Meyer L (eds) Climate change 2007: mitigation. Contribution of working group III to the fourth assessment report of the intergovernmental panel on climate change. Cambridge University Press, Cambridge
Jasanoff S (2010) A new climate for society. Theory Cult Soc 27(2–3):233–253. doi:10.1177/0263276409361497
Kirby V (2008) Natural conver(sa)tions: or, what if culture was really nature all along? In: Alaimo S, Hekman S (eds) Material feminisms. Indiana University Press, Bloomington, pp 214–236
Lambrou Y, Grazia P (2005) Gender: the missing component in the response to climate change. Food and Agriculture Organization, Rome. http://www.fao.org/sd/dim_pe1/docs/pe1_051001d1b.pdf
Latour B (2004) Why has critique run out of steam? From matters of fact to matters of concern. Crit Inq 30(2):225–248. doi:10.1086/421123
Latour B, Woolgar S (1979) Laboratory life: the social construction of scientific facts. Sage, Beverly Hills
Martin E (1991) The egg and the sperm: how science has constructed a romance based on stereotypical male–female roles. Signs 16(3):485–501
Martinez-Alier J (2003) Mining conflicts, environmental justice, and valuation. In: Agyeman J, Bullard RD, Evans B (eds) Just sustainabilities: development in an unequal world. MIT Press, Cambridge, MA, pp 201–228
McCright AM, Dunlap RE (2010) Anti-reflexivity: the American conservative movement's success in undermining climate science and policy. Theory Cult Soc 27(2–3):100–133. doi:10.1177/0263276409356001
Newman R (2001) Making environmental politics: women and love canal activism. Women's Stud Q 29(1/2):65–84
Oreskes N, Conway EM (2010) Merchants of doubt: how a handful of scientists obscured the truth on issues from tobacco smoke to global warming. Bloomsbury Press, New York
Parks BC, Roberts JT (2010) Climate change, social theory and justice. Theory Cult Soc 27 (2–3):134–166. doi:10.1177/0263276409359018
Polanyi M (1962) The republic of science: its political and economic theory. Minerva 1(1):54–73. doi:10.1007/BF01101453
Robock A (2008) 20 reasons why geoengineering may be a bad idea. Bull At Sci 64(2):14–18. doi:10.2968/064002006
Schiebinger LL (1993) Nature's body: gender in the making of modern science. Beacon, Boston
Schneider SH (1996) Geoengineering: could–or should–we do it? Clim Chang 33(3):291–302. doi:10.1007/BF00142577
Seager J (2003) Rachel carson died of breast cancer: the coming of age of feminist environmentalism. Signs J Women Cult Soc 28(3):945–972. doi:10.1086/345456
Seager J (2009) Death by degrees: taking a feminist hard look at the 2° climate policy. Kvinder Køn &Forskning 3–4:11–21
Shepard PM, Corbin-Mark C (2009) Climate justice. Environ Justice 2(4):163–166. doi:10.1089/env.2009.2402
Shiva V (1988) Staying alive: women, ecology, and development. Zed Books, London
Sims REH, Rogner H-H, Gregory K (2003) Carbon emission and mitigation cost comparisons between fossil fuel, nuclear and renewable energy resources for electricity generation. Energy Policy 31(13):1315–1326. doi:16/S0301-4215(02)00192-1
Stanley WB, Brickhouse NW (2001) Teaching sciences: the multicultural question revisited. Sci Educ 85(1):35–49
Stehr N (2001) Economy and ecology in an era of knowledge-based economies. Curr Sociol 49(1):67–90. doi:10.1177/0011392101049001005

Terry G (2009) No climate justice without gender justice: an overview of the issues. Gend Dev 17(1):5. doi:10.1080/13552070802696839

Traweek S (1988) Beamtimes and lifetimes: the world of high-energy physics. Harvard University Press, Cambridge, MA

Tschakert P (2007) Views from the vulnerable: understanding climatic and other stressors in the Sahel. Glob Environ Change 17(3–4):381–396

Tschakert P, Dietrich K (2010) Anticipatory learning for climate change adaptation and resilience. Ecol Soc 15(2):11

Wilson EA (2002) Biologically inspired feminism: response to Helen Keane and Marsha Rosengarten, 'on the biology of sexed subjects'. Aust Fem Stud 17(39):283–285. doi:10.1080/09571260220000180098

Zhang Z (1998) Towards a successful international greenhouse gas emissions trading. MPRA paper. University Library of Munich, Germany. http://econpapers.repec.org/paper/pramprapa/13023.htm

Chapter 4
Post-conventional Approaches to Gender, Climate Change and Social Justice

Karen Bell

Abstract This chapter explores the inadequacies of conventional paradigms and argues the case for a sustained effort to build post-conventional theory in relation to gender, social justice and the environment. Drawing on the work of key post-structural feminists a post-conventional framework is explored. It is argued that while significant theorising has been done in reconceptualising these issues, there is need to maintain a focus on abstract theory and to sustain the momentum of the paradigmatic shift from conventional approaches in order to effectively re-imagine the foundations of social theory. Further, if this momentum is not sustained, neglect in theory-building at the ontological level could contribute to ongoing oppression. Without reimagining and articulating a post-conventional foundation, there is a risk – by default or design – that we draw on an inappropriate and fundamentally oppressive ontology.

Keywords Gender • Climate change • Social justice • Post conventional approach • Social theory

---

Dr. Karen Bell is a social work academic at Charles Sturt University, Wagga Wagga, Australia. Her doctoral research, completed in 2009, explores the limitations of the dominant biomedical paradigm in assisted reproduction and proposes a post-conventional paradigm to validate women's epistemic agency and promote a human rights approach in health. Karen has published articles on social work theory and philosophy, field education, qualitative methodology, women's health and post-graduate student retention.

K. Bell, B.A., B.SW., Ph.D. (✉)
School of Humanities and Social Sciences, Charles Sturt University, Boorooma Street (Locked Bag 678), Wagga Wagga, NSW 2678, Australia
e-mail: kbell@csu.edu.au

M. Alston and K. Whittenbury (eds.), *Research, Action and Policy: Addressing the Gendered Impacts of Climate Change*, DOI 10.1007/978-94-007-5518-5_4,

Introduction

Anthropogenic climate change fundamentally challenges conventional paradigms based on individualistic ontology. The impacts of environmental degradation cannot be confined within the borderlines of a map or to a specific group of individuals; it affects us all in varying ways and to different degrees (Alston 2011; Dominelli 2010). The inseparability of humans and the environment has perhaps never been so difficult to deny. Yet despite the conceptual power of holistic 'earth systems thinking', partial, conventional approaches continue to dominate mainstream discourse and progress in transforming the debate has been described as 'pedestrian' (Lovelock 2010, p.22).

Much valuable work has been done to understand and assess the causes of global environmental degradation and efforts continue to raise awareness of how human activity is impacting on the environment. Likewise, efforts to document the multidimensional impact and lived experience of environmental degradation on individuals, groups and communities is ongoing. The body of literature on epistemology and methodology underpinning efforts to improve social justice outcomes, gender equity and environmental conservation is vital as are efforts to articulate alternative foundation theory at the ontological level. But why is it so important to maintain focus on the ontological foundations of gender, social justice and climate change?

If 'ontology' is taken to mean 'ways of being in the world' (Parton and O'Byrne 2000), then examining discourses of how humans exist or our 'way of being' in the world, is critical to consider theory at this foundational level. As Cuomo (1996) argues, there is a need to articulate the abstract level of theory relating to intersectional issues such as gender oppression, social injustice and environmental degradation in order to consciously map strategies for change and to transcend conventional dualist paradigms.

Without articulating and consistently advocating for post-conventional ontologies based on inter-subjectivity and interconnectedness, conventional theory and its mirrored social arrangements will persist and efforts to improve human rights, social justice and environmental outcomes will remain on the margins of dominant discourse. Global patterns of oppression will persist and even seemingly sincere efforts to 'help' or 'change' could paradoxically in fact become nothing more than methods of social control, further entrenching disadvantage (Kemp 2011). An example of how conventional approaches result in the reproduction of gender (and other) patterns of oppression is described by Alston (2011). She discusses the gendered impacts of environmental degradation and expresses her disquiet at the lack of attention paid to these gendered impacts. She argues that the local social and community consequences of climate change need to be made transparent but notes that the debate is often dominated by political, economic and masculine agendas.

If we do not continue to develop and consolidate the tools for deconstructing privilege, privilege is likely to remain effectively unexamined and perpetuated. Without redressing advantage, how can we hope to redress disadvantage? We need to consider how privilege has been constructed and what maintains hegemony as well as how to envisage a 'less oppressive ideal of humanity and its potential for constructive cohabitability' (Code 2008, p.196). Conventional ontology is inadequate in deconstructing theses issues as privilege, hierarchy and dualism are foundational, naturalised and invisible in conventional approaches; alternative tools are clearly required.

A well-articulated, post-conventional framework based on interconnectedness, embodiment and fluid subjectivity enhances our capacity for collective action for human rights and social justice outcomes. While these ideals may seem to some to be romantic, naïve or utopian, what is the alternative? If the dominance of conventional paradigms with their reliance on ontologies of mastery and competition persists, we are effectively colluding in oppression. If we tinker at the edges of change without fundamental ontological change, what are we 'sustaining'? What are we 'conserving'? What are we 'protecting'? If we do not reform theories of our way of being in the world, are we engaging in essentially superficial change to ameliorate some of the symptoms of environmental degradation?

Before further exploration of the power of post-conventional approaches, it is useful to reconsider the ways in which conventional Western paradigms have naturalised oppression and why transformation is necessary.

The Case Against Conventional Approaches

Conventional paradigms as foundational to gender oppression, social injustice and degradation of our environment, derive from ontology in which individualism, competition, hierarchies and domination are seen as the basis of existence. Western science relies on a dualistic, androcentric ordering of reality with the gender hierarchy playing a 'paradigmatic role' as a 'location of power' in producing the patriarchal order that sanctions the domination of nature and women (Braidotti et al. 2004, p.31). There is a 'systematic asymmetry' at the ontological base of conventional theory in its legitimation of white males at the top of the hierarchy. The male subject is disembodied and gains transcendence whereas female and 'other' subjects are embodied, perhaps 'over-embodied' (Braidotti et al. 2004, p.38). Thus masculine viewpoints are conflated with human viewpoints and 'others' are excluded as epistemic agents – and ontologically, out of order (Braidotti 1991). Competition rather than co-operation is central to these epistemologies of mastery (Code 1991). Thus, some humans are constructed as naturally strong and entitled to dominate and 'others' are weak and vulnerable in this 'invisible' hegemonic ontology (Haraway 2008).

The transcendence of the (male) subject conceptually disconnects humans from our environment. The environment is viewed mechanistically as a static resource, ripe for dominance, ownership and exploitation, rather than as a whole dynamic system of which humans are a part (Warren 1996). The exceptionalism of 'man' has underpinned this sense of entitlement along with the Western project to modernise post-colonial societies. However, the dominant development paradigm has not 'delivered its promise of leading every nation of the globe to the land of freedom and plenty' (Braidotti et al. 2004, pp.6 and 24); rather, it has resulted in entrenched, global patterns of social injustice, human rights violations and environmental degradation. The practice and underpinning assumptions of Western science are indeed 'important factors in the crisis of environment and development' (Braidotti et al. 2004, p.10).

Overall, the masculinist concerns of the 'global North' continue to dominate the environmental welfare debate with the basic needs of the global South at times coming a distant second in the hierarchy. Human needs often become homogenised and left at the margins, with the debate centring on how to maintain economic growth while minimising the impacts of environmental harm or on preserving what 'we' have for future generations, on 'political point-scoring' or on transforming natural resources into 'tradable commodities' (Alston 2011, p.57). This is despite the more immediate concerns of the global South generally tending to be linked to basic survival: access to clean air and water, food sovereignty and energy security (Catney and Doyle 2011). Failure to account for differences between people further disempowers oppressed people and maintains the hegemony of the oppressors.

We also need to remain mindful of how the dominant climate change agenda 'was manufactured in a crucible of inequality' by Western science and largely in the interests of the global North and we do indeed need to be 'attentive to the histories of exploitation' if we are to engage in transformative change (Cuomo 2011, pp.693, 695). At best, conventional ontology results in the dualistic global North/global South approach to climate change and new maps of oppression, disadvantage and privilege inscribed on old maps as another form of imperialism 'this time wearing a green cloak' (Catney and Doyle 2011, p.188). And at worst, conventional approaches underpin the outright denial of human-induced environmental degradation and dismiss the need for us to live within our means to prevent further degradation. To 'rip into the planet's rhythms, cycles, and interconnections, as the civilisation we have created is doing, signals human folly, not mastery' (Crist and Rinker 2010, p.13).

Conventional approaches lack the capacity to adequately theorise human interconnectedness, interdependence and our place as *part of* our planet's environment as distinct from the *owners of* it. We need to unsettle established dichotomies and further articulate a post-conventional basis for literal and figurative ecological thinking (Code 2006; Rooney 2008). If close analysis of the ontological theory base is omitted or neglected, there is indeed a real risk that these 'old maps' of oppression will simply be reinscribed on the 'new', thereby reinforcing oppression. Entirely new maps are needed in order to thoroughly reconceptualise gender, social justice and our place in our environment (Catney and Doyle 2011, p.189; Haraway 2008). The conceptualisation of human interconnectedness and interdependence is fundamental to ensure meaningful and sustainable change for social justice, equality and environmental responsibility. Efforts to consolidate a transformative reconceptualisation need to be sustained at all levels of theory and knowledge-building as a sustainable foundation for real change. Let us now explore what some key theorists offer by way of post-conventional theory.

The Case for Post-conventional Approaches

Feminist post-structural philosophers have been particularly effective in exposing how conventional Western science is inextricably linked with power, gender and knowledge as well as articulating emancipatory post-conventional paradigms based on multiple subjectivities (Braidotti et al. 2004). Warren (1996) describes the range of connections between a worldview based on dominance and social injustice, gender oppression and environmental degradation, including the historical, conceptual, experiential, symbolic, ethical, theoretical and political aspects of conventional paradigms. She reinforces the need to re-think and re-conceive the mainstay philosophical notion of dualism, to challenge conventional reason and rationality and to consolidate non male-based forms of ethics, symbolism, politics and theory.

It is clear that a fundamental transformation is required and that a superficial inclusion of those excluded into aspects of the dominant debate is inadequate. The concerns of those cast as 'other' in the conventional paradigm cannot be simply added to the dominant mix or attached at the margins. Serious engagement at the ontological level is necessary (Harding 2009) and the 'interconnections of life' must be 'obligated' into theory (Crist and Rinker 2010, p.320).

Luce Irigaray's dual subjectivity or Rosi Braidotti's nomadic subjectivity or Margrit Shildrick's fluid subjectivity or Donna Haraway's cyborg or Vandana Shiva's earth democracy could be used to articulate intersubjectivity as the basis for a post-conventional approach to gender, social justice and climate change. All of these theorists are working toward ontological reconceptualisation and recognise the need to 'truly change the subject', as Val Plumwood encouraged us to do (Plumwood 1993, p.459) in order to usurp the power of dominant discourses underpinned by dualistic hierarchies, competition and control.

For example, Luce Irigaray (1984, 1985) explores an alternative paradigm based on cooperation and exchange between human beings and rejects the dominant paradigm and its reliance on hierarchies and oppositional relations. In particular, she argues that conventional paradigms rely on disembodiment producing a mechanistic view of human beings and as such, inadequately accounts for social and other exchanges between humans. Importantly, Irigaray argues for the validation of difference, not as difference *from* the male standard subject but rather as differences *between* equal subjects.

Rosi Braidotti exposes the weaknesses of conventional paradigms, particularly in relation to the interconnectedness of human beings. Braidotti focuses on the ontological level and calls for 'nomadic' subjectivity rather than the individual, disembodied subjectivity underpinning conventional approaches (Braidotti 1994). Thus conceptualised, humans are embodied and seen to experience the world in multiple, complex and perhaps even contradictory ways. This, she argues, exposes the fundamental weaknesses and ontological frailty of conventional paradigms. She urges feminists to work for epistemological, political and ethical transformation arguing that without such paradigmatic change, the environment and 'others' will continue to be seen as disposable commodities to be exploited for maximum commercial gain (Braidotti 2005).

Dynamic ontology is conceptualised by Margrit Shildrick as ‘fluid subjectivity’ (Shildrick 1997, 2006). She too challenges conventional Western thinking and rejects the Cartesian erasure of corporeality and the consequent philosophical disadvantages it places on women and ‘others’, including the global South (Bell 2012). Shildrick also aims to deterritorialise knowledge arguing that traditional knowledge fragments human experience and often reduces our interactions to mechanistic and oversimplified universal rules. This obscures the transactional nature of human-to-human interactions as well as human-to-environment interactions and interdependence. This deterritorialisation would pave the way for multidisciplinary collaboration and for an enrichment of our knowledge of the world.

Vandana Shiva (2006) refers to ‘ancient wisdom of inseparability and interconnectedness’ and articulates a theory of earth democracy in which we are all part of a community of beings supported by the earth. Life is seen as a continuum between human and non-human species and the past, present and the future. She argues that this worldview conceptualises the earth as ‘commons’ and our inseparability from our context therefore brings with it rights as well as responsibilities. For if we are all connected and interdependent, it is necessary to respect the commons thereby creating a globalisation based on equity and collectivism rather than on self interest, domination and greed. This ontology exposes the obscenity of conventional ontology and its naturalisation of environmental, gender and economic exploitation. She argues that growth and prosperity for a few, means scarcity and poverty for many and that reclaiming ‘democratic control over food and water and our ecological survival’ is a ‘necessary project for our freedom’ (Shiva 2006, p.5).

Donna Haraway likewise aims to radically subvert the traditional individualised subject and to develop a more comprehensive, representative, inclusive and rich system of knowledge and an ‘alternative globalisation’ (Haraway 2008, p.3; Haraway 2004). As with Shiva, Haraway also presents the ontological strength of ‘becoming with others’ both human and non-human, including technologies, creatures, landscapes and practices (Haraway 2008, p.244). She also establishes the importance of situated knowledge; that is the recognition of knowledge as grounded, contextual, complex and partial. This then highlights the importance of seeking and documenting the lived experiences of phenomena and restores epistemic agency to a range of ‘actors’. As with Irigaray, Braidotti, Plumwood and Shildrick, Haraway argues that conventional ontology’s disembodied singular subject restricts the epistemic power of many human and non-human actors and that the lived experience of these ‘actors’ is devalued and invisible. This, she insists, leads to an incomplete knowledge of existence and is ‘bad science’ due to these erasures and oversights.

Another important feature of Haraway’s work is her ‘cyborg’ concept and how it is used to illustrate the interconnectedness of humans to other humans and to other non-human living things as well as to physical non-living things, such as machines and the physical environment. Theorising these interconnections is fundamental to post-conventional subject-object relations and provides the means to ‘see’ interdependence and to validate it and thus to appreciate the consequences of actions, inaction and interactions.

Thus, post-conventional ontologies are characterised by interdependent, embodied subjectivity as well as the inclusion of the lived experiences of all as equal, valid epistemic agents. This contrasts with conventional ontology and its reliance on hierarchy, dualism, exclusion and disembodiment. An axiological shift from conventional to post-conventional ontology clearly has fundamental implications for the type of knowledge (epistemology) produced and the methodology of knowledge creation. Using conventional approaches, expert, objective, deductive knowledge is valued and, by and large, the interests of the global North are dominant and maintained. Likewise, patterns of gender oppression and social injustice are naturalised and perpetuated. Post-conventional approaches offer a fuller, more equitable and sustainable approach to knowledge production as well as a focus on the interests of all beings and the environment.

Consolidating Post-conventional Approaches

Resistance to dominant conventional paradigms needs to be multilevel and likewise action to redress environmental degradation, gender inequity and social justice issues 'demands a complex process of simultaneous and interdependent transformations on different levels' (Braidotti et al. 2004, p.14). Multidisciplinary approaches to knowledge-building and interdisciplinary scholarship encourage innovative approaches to issues affecting us all on a global scale. Collective, co-operative approaches to social justice, gender equity and environmental conservation are needed for us all.

As previously discussed, conventional approaches are founded on a range of dichotomies (nature-culture, male–female, etc.) including a dichotomy between theory and practice. An important site for post-conventional theory building and action involves resistance to this false theory-practice dichotomy. Maintaining the dialecticism between theory and practice is politically and philosophically necessary to underpin real change (Cuomo 1996). The 'inextricable unity of thinking and living, theory and practice, intellectuality and politics' needs to be at the ontological, methodological and epistemological bases of post-conventional approaches (Braidotti et al. 2004). As such we should be consistent and clear in our articulation of the grounded nature of theory, the standpoint of the researcher and form democratic, co-constructed relationships with research participants (Bell 2011). This will assist us to consciously map the theory inherent in ecofeminist activism for social justice and human rights (Cuomo 1996) and guard against the epistemic imperialism of conventional approaches (Code 1991).

Standpoint research is then of central importance in the post-conventional project as it produces situated knowledge and documents the grassroots perspectives and lived experiences of under-advantaged groups and is thus fundamental in producing a 'more rationally justifiable and politically useful knowledge' for exploited groups of people (Harding 2009, p.197). Standpoint projects are well placed to challenge conventional research methodology and to avoid the illusion of expert 'objective' knowledge, or 'knowledge from nowhere' as Haraway calls it (Haraway 2004). Multidisciplinary standpoint projects also disrupt the reproduction of social power inherent in conventional approaches to research in situating all humans as epistemic agents (Harding 2009). As conventional approaches have effectively silenced all sorts of individuals, groups and communities, it is imperative to resist such 'epistemologies of ignorance' and ask 'who benefits and who is disadvantaged by such ignorance?' (Tuana 2004, p.196).

Multidisciplinary projects also reflect a more holistic, interconnected and interdependent approach to ontology, epistemology and methodology. As such, the 'territorial' nature of conventional knowledge is challenged and a more comprehensive, inclusive form of knowledge is produced (Shildrick 2006; Lovelock 2010). With the restoration of epistemic agency to previously disempowered people, the exploration and documentation of their lived experience is not only validated but is central in building a comprehensive, inclusive knowledge base as a foundation for further action and theory-building, policy and action.

Conclusion

Thus it is imperative that efforts are ongoing to consciously re-envision foundation theory and consolidate a post-conventional paradigm that adequately theorises subjectivity in its complexity, rather than conventional subjectivity and its oppressive, simplistic and hierarchical reductionism. A fluid/nomadic/dual/cyborg ontological base opens up rich conceptual possibilities to better capture the interdependence of humans and our environment. Such a post-conventional ontology restores epistemic agency to oppressed people and recognises equity, co-operation and collectivism as think-able, do-able and desirable. If we fundamentally believe that we are all connected and that we exist as part of the planet's ecosystem, then others' wellbeing impacts on our own wellbeing; our actions in the world are recognised as impacting on others. And if collectivism is fundamental to existence, then taking only one's fair share of our environment's resources is a given. Existence based on sustainability, co-operation and fairness would expose the folly of conventional, ceaseless efforts to 'grow', compete and dominate.

References

Alston M (2011) Gender and climate change in Australia. J Sociol 47(1):53–70

Bell K (2012) Towards a post-conventional philosophical base for social work. Br J Soc Work 42(3):408–423

Bell K (2011) Participants' motivations and co-construction of the qualitative research process. Qual Soc Work. doi:DOI 10.1177/1473325011429020

Braidotti R (1991) Patterns of dissonance: a study of women in contemporary philosophy. Polity Press, Cambridge

Braidotti R (1994) Nomadic subjects: embodiment and sexual difference in contemporary feminist theory. Columbia University Press, New York

Braidotti R (2005) A critical cartography of feminist postmodernism. Aust Fem Stud 20(47): 169–180

Braidotti R, Charkiewicz E, Hausler S, Wieringa S (2004) Women, the environment and sustainable development: towards a theoretical synthesis. Zed Books, London

Catney P, Doyle T (2011) The welfare of now and the green (post) politics of the future. Crit Soc Policy 31(2):174–193

Code L (1991) What can she know? Feminist theory and the construction of knowledge. Cornell University Press, New York

Code L (2006) Ecological thinking: the politics of epistemic location. Oxford University Press, Oxford

Code L (2008) Thinking about ecological thinking. Hypatia 23(1):187–203

Crist E, Rinker H (eds) (2010) Gaia in turmoil: climate change, biodepletion, and earth ethics in an age of crisis. Massachusetts Institute of Technology, Cambridge, MA

Cuomo C (1996) Toward thoughtful ecofeminist activism. In: Warren K (ed) Ecological feminist philosophies. Indiana University Press, Bloomington

Cuomo C (2011) Climate change, vulnerability, and responsibility. Hypatia 26(4):690–714

Dominelli L (2010) Social work in a globalizing world. Polity Press, Cambridge

Haraway D (2004) The Haraway reader. Routledge, New York

Haraway D (2008) When species meet. University of Minnesota Press, Minneapolis

Harding S (2009) Standpoint theories: productively controversial. Hypatia 24(4):192–200

Irigaray L (1984) An ethics of sexual difference. Cornell University Press, New York

Irigaray L (1985) Speculum of the other woman. Cornell University Press, New York

Kemp S (2011) Recentring environment in social work practice: necessity, opportunity, challenge. Br J Soc Work. doi:DOI.10.1093/bjsw/bcr119

Lovelock J (2010) Our sustainable retreat. In: Crist E, Rinker H (eds) Gaia in turmoil: climate change, biodepletion, and earth ethics in an age of crisis. Massachusetts Institute of Technology, Cambridge, MA

Parton N, O'Byrne P (2000) Constructive social work: towards a new practice. St Martin's Press, New York

Plumwood V (1993) The politics of reason: towards a feminist logic. Aust J Philos 71(4):436–462

Rooney P (2008) Epistemic responsibility and ecological thinking. Hypatia 23(1):170–176

Shildrick M (1997) Leaky bodies and boundaries: feminism, postmodernism and (bio)ethics. Routledge, London

Shildrick M (2006) Monstrous reflections on the mirror of the self-same. In: Orr D, Lopez-McAlister L, Kahl E (eds) Belief, bodies and being: feminist reflections on embodiment. Lanham Rowman and Littlefield, New York

Shiva V (2006) Earth democracy: justice, sustainability and peace. Zed Books, London

Tuana N (2004) Coming to understand: orgasm and the epistemology of ignorance. Hypatia 19(1):194–232

Warren K (1996) Ecological feminist philosophies. Indiana University Press, Bloomington

Chapter 5
Two Solitudes, Many Bridges, Big Tent: Women's Leadership in Climate and Disaster Risk Reduction

Elaine Enarson

Abstract Despite commonalities in the theoretical, policy, and practical domains of climate and disaster work, unnecessary divisions persist. The chapter posits that gender analysis, too, overlooks important synergies and replicates the unhelpful 'two solitudes' approach. The discussion then turns to identifying positive models and concrete steps for bridging these gaps. Given the integral relationship on the ground between gender, climate and disaster, a 'big tent' approach is urged to reflect the concerns, resources, and expertise of gender, climate, and disaster actors equally. Neither disaster risk reduction nor climate adaptation is women's work alone, but the historic organizing of women for social justice positions them as leaders toward community resilience.

Keywords Disaster • Gender relations • Climate change • Environmental challenges • Women's leadership

Disaster reduction and climate adaptation efforts around the world continue to develop along parallel paths, as unnecessary and unwise as this is. The lines that divide these communities of practice run deep, and both discourses too often remain divorced from the global development processes that create risk and vulnerability. The divergent realms may now be too deeply entrenched but this cannot be said about gender, disaster and climate work. In fact, enabling women's leadership in both domains is potentially the most positive and unifying force in our era, when an integrated approach is essential. Examining writing about gender and disaster risk from the global South and North, Lourdes Meyreles and I (Enarson and Meyreles 2004) described "parallel paths" in critical thinking and practical action. With climate change factored in, an approach even more fragmented is not

---

E. Enarson, Ph.D. (✉)
Independent Scholar, Boulder, Colorado, USA
e-mail: enarsone@gmail.com

M. Alston and K. Whittenbury (eds.), *Research, Action and Policy: Addressing the Gendered Impacts of Climate Change*, DOI 10.1007/978-94-007-5518-5_5,

viable. The issues are "joined at the hip" with no meaningful theoretical or practical daylight supporting claims for parallel universes. If ever we had this luxury, we no longer do.

Below, I briefly explain why disaster and climate risk reduction are so distinct and yet so related (the two solitudes), before discussing women and gender relations as cross-cutting currents, sometimes replicating traditional fracture lines but also demonstrating commonalities (the bridges). Here I question the apparent but superficial convergence of gender, climate and disaster work. The chapter concludes with suggestions for meeting women where they are, especially community women already mobilizing to meet environmental stress and the many other challenges to their livelihood, health, safety and autonomy. With women's empowerment as a central goal, and a generous, inclusive, and non-dogmatic approach (the big tent), the future seems less bleak.

The Two Solitudes: Difference and Commonality

Climate change cannot be reduced to a series of disasters but neither are disasters a footnote in the story of a warming planet. Others have sketched out the differences, including diverse worldviews, terminology, professional networks, institutional frameworks, research agendas, practical goals and political positions (O'Brien et al. 2006; Schipper and Pelling 2006; Kelman and Gaillard 2008). If not as entrenched as the original "solitudes" of Francophone and English Canada, these are critical divisions.

Disaster and climate experts often fail to communicate (to say nothing of communication gaps with community members) giving a different nuance to "mitigation" and "adaptation," and working from different literatures. The knowledge gained by those who study disasters do not figure large in the reports of climate experts, though increasing attention is apparent in more recent reports of the Intergovernmental Panel on Climate Change in public health, risk perception and communication, and culturally-specific adaptation strategies, among other concerns held in common with disaster specialists. Climate research is dominated by the "hard" sciences if not drifting in the "hard wired" direction as geo-engineering approaches gain popularity. In contrast, the area study of disaster attracts a wider range of analysis from the human and social sciences, as well as from natural scientists, engineers, land use planners, and those in business management and public administration. Disciplinary blinders are nothing new, but in the rapidly evolving realm of climate risk it seems short-sighted not to define the subject as generously as possible.

Neither climate science nor disaster science lays sole claim to the critical knowledge needed to turn the ship around. Focusing more on adaptation than on prevention, and more on livelihood in the global South, climate change experts bring increased awareness of the insidious changes associated more with a warming climate than with sudden-onset disasters, but also examine weather-related disasters, slow-motion environmental displacement, and the geo-politics of involuntary migration. In the disaster field, the focus is more on events than on process. Though still a contested term, "disasters" are understood as fundamentally sociopolitical events with damaging material effects that transcend local coping capacity (Perry and Quarantelli 2005). Often seen as bounded ("before, during and after disaster") and cyclical (mitigation, preparedness, relief, recovery), the antecedent causes are often apparent, as when a nearby volcano comes to life.

Students of disaster also take an all-hazard approach. Whether triggered by an unexpectedly large but predictable seasonal flood, the breach of a poorly designed, sited or maintained dam, or water displaced through earthquake or landslide, a flood is a flood is a flood. In this way, disaster researchers may miss the forest for the trees, seeing a flood or drought as more isolated than it is and missing the larger social processes and decisions that produced it. This short-hand portrait minimizes the critical contributions of disaster researchers who strive to identify the historical and contemporary drivers magnifying hazards and social vulnerabilities while undermining people's capacity to anticipate, plan for, resist, and respond (Wisner et al. 2004), but the field as a whole still lists in a more conservative direction.

Like their colleagues working on climate, disaster researchers have written a wide-ranging social science building on the foundational knowledge of natural scientists and, to a lesser degree, on indigenous knowledge and practice in communities at risk. The cannon has been elaborated in different cultural and hazard contexts and time periods (in a crowded field, see the research summaries offered by Tierney et al. 2001; National Research Council 2006; Rodríguez et al. 2007; Haque 2006; Pelling 2003). We know a fair amount about how people in different social, geographic, and cultural locations relate to natural phenomena, about the "everyday disasters" that trigger social crisis, and about how people strive to make sense of untoward events.

We have come to understand the social process of producing, sharing, assessing and acting on warnings, forecasts, and other information about risk, and the new communication media that facilitate (or may hinder) this. The specific factors shaping preparedness and evacuation decisions, emergency response efforts, and long-term relief are well established, and those promoting individual and community resilience increasingly coming under examination. We have certainly learned about the social construction of "disasters by design" (Mileti 1999). Through qualitative case study data and sophisticated quantitative modeling, the fundamentals of social vulnerability and resilience have been mapped theoretically and empirically, so the disaster community also has a good sense of how disaster risk can be reduced, the measurable costs and benefits of differing models of mitigation, and how sustainable recovery can be promoted.

Many common concerns should bring these two solitudes into closer orbit. The high costs of a steadily warming climate are manifestly borne by the world's poorest in the world's poorest nations, as are the effects of climate disasters. Short- and long-term displacement is a common concern, as are the economics of environmental uncertainty. Like disasters, the effects of climate variability are not one-dimensional or universally negative; also like disasters, the effects of climate change are endured mainly by those far removed from the antecedent individual, corporate and governmental actions and inactions that produced them. Uncertainties are at the heart of both, and so are the vast social inequalities complicating people's capacities to positively cope with environmental stress. Both climate change and disasters are quintessentially human constructions; not climate, but the way we have built our livelihoods and settlements on the planet historically is at issue, and the vulnerabilities arising from dominant productive process, power relations, and patterns of governance.

Finally, gender is a common theme cutting across climate and disaster discourse. The "gendered terrain" of disasters illuminated over the past two decades (Enarson and Morrow 1998; Siriner et al. 2010; Enarson 2009a) adds depth to earlier work around dispossession and marginalization, bringing environmental change and extreme weather events into focus through the eyes of women and men, boys and girls. As contributing authors attest, gender analysis is increasingly salient in climate work as it is in disaster work globally (Dankelman 2011; Röhr et al. 2009; UN DAW 2001).

Gender, Climate and Disaster: Degrees of Separation

Very unfortunately, fracture lines more than commonalities are evident in feminist or gender-focused initiatives around disaster and climate, as in the wider discourse. An integrated approach might be inferred based on titles, photo illustrations, and occasional cross-referencing but, in my view, these synergies are more apparent than real. In fact, gender-focused climate and disaster researchers, policy makers, practitioners, and activists operate on different planes and sometimes at cross-purposes.

Theory and Research

Gender and disaster researchers analyze post-event shifts in gender power relations and a host of other issues such as sustainable livelihoods, cultural survival, access to information, capacity to prepare and seek safety outside the home when necessary, gender violence, threats to women's land rights and other human rights, reproductive health, migration, housing, displacement, family dynamics, women's collective work in crises, and the gender dimensions of emergency relief and recovery. While less is known about men and masculinities, about sexualities generally and stigmatized sexual minorities specifically, gender is widely recognized as relevant in virtually every corner of the disaster library.

These are pressing concerns for most gender and climate researchers, too, but knowledge exchange is not the norm, for example in journals where readers will find two special issues on gender and climate (Terry 2009; Masika 2002) and two on gender and disaster (Phillips and Morrow 2009; and see the 2010 collection edited by Houghton). Climate and gender specialists rarely draw on the new gender and disaster subfield; by the same token, students of disaster rarely examine shifts or tensions in gender relations related to changing climatological conditions. In my own work, for instance, only in a short study from Gujarat, India after the 2001 earthquake did I examine antecedent conditions (long-standing drought) and then without considering the drivers of this drought or its implications for disaster resilience (Enarson 2001; and see case studies in Enarson and Chakrabarti 2009). Recent theoretical critiques written from a climate change perspective promise to help write more theoretically nuanced accounts that highlight intersectionality, contradiction, and subjectivity (McGregor 2010; see Fordham 2004), rejecting the once-dominant vulnerability paradigm, challenging false dualities of "virtue or vulnerability," and not glossing over contradictions and counter-intuitive evidence about women in different parts of the world (Arora-Jonsson 2011).

Parallel paths of theory and research blind us to the convergence of forces in the everyday worlds of the women and men, boys and girls with whom we are concerned as researchers. Competition for research grants and related resources is inevitable in both camps, and worse in areas of overlap. The framing of concerns theoretically shifts discourse (and the attention of decision makers) from one extreme to another, asking the wrong question: Climate or disaster-what's the real problem? A virtue in the abstract perhaps, but competing or parallel approaches ultimately limit mutuality and cooperation just when theoretically-informed practical action is essential in the face of common risks.

Policy, Practice, and Advocacy

Gender-focused reports and guidelines from humanitarian relief agencies responding to emergencies are generally earmarked "disaster" or "climate" but rarely both. Gender mainstreaming is clearly underway in both domains, though differently. UN agencies engaged in humanitarian relief, gender, environmental, and disaster work strive to bring gender sensitivity home while lead agencies promote mainstreaming tools and strategies focused on risk reduction for others (e.g. ISDR 2009). The website of UN Women, the newly consolidated family of UN agencies working toward gender equality, highlights climate with only a nod to (climate-related) disasters. Similarly, gender mainstreaming guides geared to disaster practitioners, tend to introduce climate-related hazards and vulnerabilities through field material (a text box) or in a single chapter, but just as often these are neglected altogether (for two among many examples, see Enarson 2009b; Mehta 2007).

Ostensibly integrated campaigns, such as the UNDP's (2009) work in the Caribbean on "enhancing gender visibility in disaster risk management and climate change," actually continue to segregate gender-focused climate and disaster work. Separate tables on climate and disaster, in Oxfam's training materials geared to synthesis, are another small example of how rarely in practice gender, climate and disaster are taken up jointly (Oxfam 2010). Few virtual climate-related networks (for example, 350.org) and even fewer based in the academy reach out to women and men who routinely face non-climate disasters that undermine capacities, exacerbating vulnerability to climate uncertainty.

At international policy tables, neither camp holds much weight; both communities of practice struggle to speak and be heard, diverting resources and sapping spirit for campaigns conceived more broadly that might bring the "two solitudes" together in common cause. Climate-centered National Adaptation Programmes of Action rarely prioritize disaster reduction or connect the dots between gender equality, disaster risk, and positive adaptation to climate change. Gender has more salience in the language of key disaster risk policy documents such as the Hyogo Framework for Action, where gender is recognized as a key cross-cutting principle, but not in relation to climate (Enarson 2009a; GDN 2009). Global coalitions such as the Global Gender and Climate Alliance, engaging UNDP, UNEP, WEDO and IUCN, have no real parallel in the gender and disaster field, and perhaps for this reason gender activists organized around water, energy, forestry, food security, land rights, peace keeping and related issues come more readily to climate adaptation than to disaster risk reduction as their central concern. The 2008 Manila Call for Action, arising from a global conference attracting high-level participants in both domains, helped raise the profile of women's potential leadership in both areas. Despite these and other shifts in discourse, alliances between the two gender constituencies seem narrow and fleeting. As in the broader climate and disaster risk communities, this undermines the potential for women's effective collaboration on specific, place-based campaigns that respond to women and their families on the front lines, for whom these "lines that divide" are artifacts of the imagination.

Bridge Building

Despite these cautionary notes, ample grounds exist for optimism that gender analysis can bring much-needed synergy to the enormous environmental, social, and political challenges facing women and men across the planet.

With respect to research, if most of us labor in our own fields, outstanding examples can be found of gender, disaster, and climate writing with a broad brush in an integrative way (among others, see Dankelman 2010, on climate, disaster and human security). Overlapping questions and answers are now too important to ignore (see reviews by Brody et al. 2008; Fothergill 1998; and see Enarson 2012). The literature on disparate impacts, including gendered mortality and morbidity, is well-developed. Parallel gender gaps are found in women's higher support for mitigation and positive adaptation (in climate research) and for mitigation and preparedness (in disaster research). Challenges to traditional gender norms in the face of environmental stress, due to climate or disaster or both, are commonly reported, and women's caregiving responsibilities expand in crises regardless of root causes. Both climate and disaster researchers with a gender lens document shifts in the gendered division of labor as families adapt to drought, for instance, or are forced to relocate after an earthquake. Women's and men's uses of media have also been explored from both perspectives. Significantly, feminist research designs are also common, drawing out the stories of indigenous women and other in Northern

Canada to the Pacific Islands and beyond. Happily, both literatures are now too large to quickly summarize (see resources on the websites of GenCC and the Gender and Disaster Network).

With respect to training and practice, promising practices are reported in accounts of projects based on gender equality and disaster/climate risk reduction (ISDR 2008). Some more than others make a concerted effort to introduce disaster into climate work (see Aguilar 2009, for a positive example). While the World Bank highlights "three things you should know" (2011) about gender and climate change—disaster risk reduction among them—benign neglect between the "two (gender) solitudes" is more the norm. Virtual networking may be a stronger platform for collaboration in future. Certainly, the growing number of gender, climate and/or disaster networks and websites foreshadow connections that begin with sharing resources and cross-linking websites, but go further. Financial exigencies may drive a collaborative and integrated approach where gender is concerned, too.

A supportive social infrastructure is also evident, for instance in the Gender and Water Alliance and the stellar work of the IUCN in both domains. The UNDP's Bureau of Crisis Prevention and Response is working toward a global action research center to build knowledge and capacity around climate change and disaster, along with armed conflict, gender-based violence, governance, and HIV/AIDS, illustrating the positive value of a holistic approach.

Finally, encouraging trends toward integration can be seen in local community work, where gender sensitive campaigns can be found that raise the profile of environmental risk without pre-determined divisions. The community resilience campaign of Grassroots Organizations Operating Together in Sisterhood (GROOTS), a coalition of dozens of community based women's groups around the globe, is also heartening. After nearly two decades of on-the-ground collaboration with women and their families trapped in hazardous living conditions and highly exposed to disaster, the network smoothly integrated climate risk reduction when its significance became known. The India-based Self-Employed Women's Association (SEWA) is a well-known instance of proactive organizing that marries hazard mitigation (regardless of specific triggering factors) with capacity development to protect and enhance women's rights (e.g. seismically resistant masonry, leadership courses, environmental education, disaster insurance through women's banks). Numerous other examples of small-scale projects conceived with attention to climate and disaster risk alike can be found in recent "good practice" documents (all warranting more analysis).

More encouraging yet are the emergent women's movements that arise in the wake of specific disasters and specific threats to livelihood. Women call for respecting and promoting human rights in the aftermath (see Burnad 2006, for one example from the Indian Ocean tsunami; and Enarson and Fordham 2001). The politics of place are compelling when the land, water, air and sea that ground a people come under assault, and this is often an entry point for women's environmental activism. Women have mobilized in large numbers—not universally or automatically, but persuasively—when land, livelihood, and health are jeopardized, whether due to erosion or floodwaters, an earthquake or extreme heat, knowing that cultural survival may be at stake, too. The social bonds that unite them in defense of home and community are there to see, bringing something essential to the larger political movements around reducing disaster and/or climate risk (among others, David and Enarson 2012; Enarson and Chakrabarti 2009; Gender Focal Point 2008; Yonder et al. 2005; Rocheleau et al. 1996).

Big Tents at the Crossroads: Gender Justice and Environmental Risk Reduction

Internal pressures to the contrary, it is as urgent as ever to move from gender analysis to feminist movement—and from the progressive efforts of men and women together, as important as these are, to organizing women and girls specifically as "keys to prevention." This claim, made by disaster specialists in 1995, is no less true today and applies as well to climate change adaptation. Nothing inherent to the female sex, but much about contemporary gender relations moves women toward a holistic and action-oriented approach to environmental problems.

Women's leadership toward positive adaptation and disaster mitigation and preparedness is well-documented (Brody et al. 2008; Carvajal-Escobar et al. 2008). But it is important not to overstate the case: the cartography of danger is highly gendered. Women and men in different social locations are not equally able to marshal or control resources of survival in environmental crisis, or take protective action to forestall crisis. Women and men produce everyday life—the materiality of it and its symbolic meaning—very differently. Gender relations bring women's and men's lives into close alignment but are also divisive and can be destructive, and stubborn (Eriksen et al. 2010; Bradshaw 2001). Conflicting or overlapping gender interests around the use of forests and arable land, around water, and around land rights are all too apparent, both in the wake of natural disasters and in debates around reducing emissions, best illustrated recently in women's critique of the Reduced Emissions from Deforestation and Degradation (REDD) framework in climate negotiations. Gender interacts with other power dynamics in work cultures, obviously including environmental organizations and government agencies, as it does with respect to mitigation and adaptation, preparedness, evacuation, migration, and a host of other critical decisions that families and neighbors must make in the face of environmental crisis.

The question is complex, but the short-hand conclusion from academia and from relief and development workers is that women's energy is critical to successful local action. This is true in wealthy and in poor countries, with Wangari Matthei's Green Belt movement just one among many stellar examples, and consider the activism of low-income women along the US Gulf Coast around sustainability and environmental restoration after hurricane Katrina (David and Enarson 2012). Women's activism may be grounded in faith or in ecofeminist values, arise from a simple ethic of conservation and care, or reflect years of peace work. For most women, livelihood protection is the essential driver, but women are also moved to action by threats to the well-being of their family, their sense of place, and their land and property rights.

The environmental justice movement is notable for uniting women and men on racial and class grounds, but the third leg is gender justice. In the US, for instance, women of color already mobilizing around reproductive rights now campaign for "looking both ways" to identify threats to women's reproductive health and rights arising from climate change and climate disasters (Rojas-Cheatham et al. 2009). Progressive women's organizing around economic and racial justice, democratic governance, and the rights of children, elders, and persons living with disabilities or HIV/AIDS offer more potential grounds for collaboration, as does antiviolence organizing in the wake of disasters (Fisher 2010; Brown 2012) and documentation projects undertaken by women affected by disaster or climate uncertainty (Gender Focal Point 2008; Anderson 2009, in the Pacific Islands). These are strong entry points for working with women not wedded to either climate or disaster discourse, but potentially active in reducing the risk of both.

As essential as men's partnership with women to reduce these risks is, at this juncture women's work is critical. Women are positioned as (behind-the-scenes) leaders in environmental risk reduction based on demonstrated passion, expertise, and local knowledge—whether they are building green energy projects, mobilizing against violence of all kinds, sharing debt to grow small home enterprises, seeking political leadership roles, educating children about fire and other hazards, or collectively managing natural resources to produce the staples of life. Without their full and active participation in reducing vulnerabilities and building resilience it seems unlikely that communities can fully prepare for or defend against a future dominated by environmental uncertainty and extremes.

Conclusions and Next Steps

The present "solitudes" in mainstream gender and climate work are unsustainable, and do not serve women well. Like men, women are directly and indirectly affected by both and care about both. Bridgework has been the hallmark of women's political mobilization in the past, and with a "big tent" approach that takes environmental disaster and climate risk to women where they are, these divisive rifts can begin to mend.

It remains to forge the connections, taking a "big tent" approach and seeking space for women's common work against common threats. Women's coordinated efforts can be the weight-bearing center beam of resilience—but linkages must be real, supported materially, and sustained in practice. Among other steps, opportunities for demonstration projects to trial, analyze, and improve local initiatives must be found, and regional action research collaboratives forged. Cross-training (gender, climate, disaster) must become the norm. Systematic joint campaigns on selected issues should be our default mode, and arise from the strong cross-cutting personal, professional, and intellectual networks we can build. Women's progressive mobilization around environmental concerns is a critical precondition (necessary but not sufficient) for moving where we need to go in the time remaining. This is not women's work alone. Further analysis is needed of how men and women, boys and girls positively engage around climate and disaster risk reduction in different contexts, and more financial and material support for trialing and scaling up coordinated and inclusive approaches.

Global women's networks, UN agencies with claims to gender leadership, and gender-aware NGOs and community organizations are all important platforms for action, but the homegrown activism of local women "at the crossroads" of environmental risks is the indispensable starting point. We cannot be a resilient people on a fragile planet without this.

References

Aguilar L (2009) Training manual on gender and climate change. International Union for Conservation of Nature and UN Development Program. Retrieved from http://cmsdata.iucn.org/downloads/eng_version_web_final_1.pdf

Anderson C (2009) Stories from the Pacific: the gendered dimensions of disaster risk management and adaptation to climate change. AusAID and UNDP Pacific Center. Retrieved from http://www.undppc.org.fj/_resources/article/files/UNDP%20PC%20Climate%20Change.pdf

Arora-Jonsson S (2011) Virtue and vulnerability: discourses on women, gender and climate change. Glob Environ Chang. doi:10.1016/j.gloenvcha.2011.01.005

Bradshaw S (2001) Dangerous liaisons: women, men and hurricane Mitch. Fundacion Puntos de Encuentro, Managua

Brody A, Demetriades J, Esplen E (2008) Gender and climate change: mapping the linkages – a scoping study on knowledge and gaps. Retrieved from http://www.bridge.ids.ac.uk/reports/Climate_Change_DFID.pdf

Brown B (2012) Battered women's shelters in New Orleans: recovery and transformation. In: David E, Enarson E (eds) The women of Katrina. University of Vanderbilt Press, Vanderbilt, pp 179–189

Burnad F (2006) Tsunami aftermath: violations of human rights of Dalit women, Tamil Nadu, India. Asia Pacific Forum on Women, Law and Development. Retrieved from http://www.apwld.org/pdf/Tsunami_India.pdf

Carvajal-Escobar Y, Quintero-Angel M, García-Vargas M (2008) Women's role in adapting to climate change and variability. Adv Geosci 14:277–280

Dankleman I (2010) Climate change, human security and gender. In: Dankleman I (ed) Gender and climate change: an introduction. Earthscan, London, pp 55–72

Dankelman I (ed) (2011) Women and climate change: an introduction. Earthscan, London

David E, Enarson E (eds) (2012) The women of Katrina: how gender, race and class matter in an American disaster. Vanderbilt University Press, Vanderbilt

Enarson E (2001) 'We want work:' Gujarat women in drought and earthquake. Quick Response research report #135, University of Colorado Natural Hazards Center. Retrieved from http://www.colorado.edu/hazards/research/qr/qr135/qr135.html

Enarson E (2009a) Gendering disaster risk reduction: 54 steps from words to action. In: Enarson E, Chakrabarti PGD (eds) Women, gender and disaster. Sage Publications, Delhi, pp 320–337

Enarson E (2009b) Gender mainstreaming in emergency management: a training manual for emergency planners. Prairie Women's Health Center of Excellence, Winnipeg. Retrieved from http://www.gdnonline.org/resources/GEM_MainFINAL.pdf

Enarson E (2012) Women confronting natural disaster: from vulnerability to resilience. Lynne Rienner Publications, Boulder

Enarson E, Chakrabarti PGD (eds) (2009) Women, gender and disaster: global issues and initiatives. Sage Publications, Delhi

Enarson E, Fordham M (2001) From women's needs to women's rights in disasters. Environ Hazards 3:133–136

Enarson E, Meyreles L (2004) International perspectives on gender and disaster: differences and possibilities. Int J Sociol Soc Policy 24(10/11):49–93

Enarson E, Morrow BH (1998) The gendered terrain of disaster: through women's eyes. Greenwood Publications, Westport

Eriksen C, Gill N, Head L (2010) The gendered dimensions of bushfire in changing rural landscapes in Australia. J Rural Stud 26:332–342

Fisher S (2010) Sri Lankan women's organizations responding to post-tsunami violence. In: Enarson E, Chakrabarti PGD (eds) Women, gender and disaster. Sage Publications, Delhi, pp 233–249

Fordham M (2004) Gendering vulnerability analysis: towards a more nuanced approach. In: Bankoff G, Frerks G, Hillhorst D (eds) Mapping vulnerability: disasters, development, and people. Earthscan, London, pp 174–182

Fothergill A (1998) The neglect of gender in disaster work: an overview of the literature. In: Enarson E, Morrow BH (eds) The gendered terrain of disaster: through women's eyes. Greenwood Publications, Westport, pp 11–25

Gender and Disaster Network (2009) The disaster risk reduction process: a gender perspective. Contribution to the 2009 ISDR Global Assessment Report on Disaster Risk Reduction. Retrieved from http://www.fire.uni-freiburg.de/Manag/gender%20docs/UNISDR-Summary-of-Gender-&-DRR-worldwide.pdf

Gender Focal Point Network (2008) Women in times of disaster, gender and disaster risk reduction: perspective from Japan. Asia Pacific Economic Cooperation. Retrieved from http://www.onlinewomeninpolitics.org/sourcebook_files/Resources/Report-%20Gender%20and%20Disaster%20Risk%20Reduction%20-%20Perspective%20from%20Japan.pdf

Haque CE (2006) Mitigation of natural hazards and disasters: international perspectives. Springer, Dordrecht

Houghton R (2010) Women and disaster. Special issue. Int J Mass Emerg Disasters 28(2):148–293

Kelman I, Gaillard JC (2008) Placing climate change within disaster risk reduction. Disaster Adv 1(3):3–5

Manila Declaration for Global Action on Gender in Climate Change and Disaster Risk Reduction Manila, Philippines (2008) 22 October Retrieved from 2008 http://www.unisdr.org/files/8731_maniladeclarationforglobalactionongenderinclimatechangeanddisasterriskreduction1.pdf

Masika R (ed) (2002) Gender, development and climate change. Special issue. Gend Dev J 10(2). Oxfam, London

McGregor S (2010) 'Gender and climate change': from impacts to discourses. J Indian Ocean Reg 6(2):223–238

Mehta M (2007) Gender matters: lessons for disaster risk reduction in South Asia. International Centre for Integrated Mountain Development, Kathmandu. Retrieved from http://www.icimod.org/?opg=949&q=drr_document&document=1082

Mileti D (1999) Disasters by design: a reassessment of natural hazards in the United States. Joseph Henry Press, Washington, DC

National Research Council, Committee on Disaster Research in the Social Sciences (2006) Facing hazards and disasters: understanding human dimensions. Retrieved from http://www.nap.edu/catalog.php?record_id=11671

O'Brien G, O'Keefe P, Rose J, Wisner B (2006) Climate change and disaster management. Disasters 30(1):64–80

Oxfam (2010) Gender, disaster risk reduction, and climate change adaptation: a learning companion. Retrieved from http://www.gdnonline.org/resources/OxfamGender&ARR.pdf

Pelling M (ed) (2003) Natural disasters and development in a globalizing world. Routledge, New York

Perry R, Quarantelli EL (eds) (2005) What is a disaster? New answers to old questions. Xlibris, Philadelphia

Phillips B, Morrow BH (2009) Women in disaster: From theory to practice. Bloomington, IN

Rocheleau D, Thomas-Slayter B, Wangarai E (eds) (1996) Feminist political ecology: global issues and local experiences. Routledge, New York

Rodríguez H, Quarantelli EL, Dynes R (eds) (2007) Handbook of disaster research. Routledge, New York

Röhr U, Hemmati M, Lambrou Y (2009) Towards gender equality in climate change policy: challenges and perspectives for the future. In: Enarson E, Chakrabarti PGD (eds) Women, gender and disaster. Sage Publications, Delhi, pp 289–303

Rojas-Cheatham A, Ginn Paredes D, Griffin S, Shah A, Shen E (2009) Looking both ways: women's lives at the crossroads of reproductive justice and climate justice. Asian Communities for Reproductive Justice. Retrieved from http://reproductivejustice.org/assets/docs/ACRJ-MS5-Looking-Both-Ways.pdf

Schipper L, Pelling M (2006) Disaster risk, climate change and international development: scope for, and challenges to, integration. Disasters 30(1):19–38

Siriner I, Dasgupta S, De Sarathi P (eds) (2010) Women's encounter with disaster. Frontpage, Kalkuta

Terry G (ed) (2009) Climate change and gender justice. Special issue. Gend Dev J 17(1). Oxfam, London

The World Bank (2011) Gender and climate change: three things you should know. Retrieved from http://siteresources.worldbank.org/EXTSOCIALDEVELOPMENT/Resources/244362-1232059926563/5747581-1239131985528/5999762-1321989469080/Gender-Climate-Change.pdf

Tierney K, Lindell M, Perry R (2001) Facing the unexpected: disaster preparedness and response in the United States. Joseph Henry Press, Washington, DC

UN Development Program (2009) Enhancing gender visibility in disaster risk management and climate change in the Caribbean. Retrieved from http://www.undp.org.cu/crmi/genderstudy/index.asp

UN Division for the Advancement of Women (2001) Environmental management and the mitigation of natural disasters: a gender perspective. Expert meeting. Proceedings retrieved from http://www.un.org/womenwatch/daw/csw/env_manage/

UN International Strategy for Disaster Reduction (2008) Gender perspectives: integrating disaster risk reduction into climate change adaptation, good practices and lessons learned. Retrieved from http://www.unisdr.org/eng/about_isdr/isdr-publications/17-Gender_Perspectives_Integrating_DRR_CC/Gender_Perspectives_Integrating_DRR_CC_Good%20Practices.pdf

UN International Strategy for Disaster Reduction (2009) Making disaster risk reduction gender sensitive: policy and practical guidelines. Retrieved from http://www.preventionweb.net/files/9922_MakingDisasterRiskReductionGenderSe.pdf

Wisner B, Cannon T, Blaikie P, Davis I (2004) At risk: natural hazards, people's vulnerability and disasters. Routledge, London

Yonder A, Akcar S, Gopalan P (2005) Women's participation in disaster relief and recovery. SEEDS pamphlet series, #22. Retrieved from http://www.disasterwatch.net/Brief/Seeds2005final.pdf

Part III
Interrogating Policy from a Gender Perspective

Chapter 6
Gendering Climate Change: Implications for Debates, Policies and Practices

Lena Dominelli

Abstract In this article, I take the view that all disasters are interconnected, and that the deleterious nature of their impact is exacerbated for marginalized peoples, including indigenous people, children, women, older people, disabled people and those living in poverty. However, this consideration is not the main focus of this chapter. Instead, I consider gendered relations within climate change and disaster debates more generally to highlight the neglect of women's experiences during such calamities and make suggestions for its rectification in policy debates and practice. In doing this, I consider the experiences of women as recipients of aid and as relief workers to reveal that gendered relations silence the specific experiences of women. I draw largely on an Economic and Social Sciences Research Council (ESRC) funded project on humanitarian aid delivered during the 2004 Indian Ocean Tsunami in one of the worst affected countries in South East Asia, Sri Lanka, but keep the locations and identity of those involved anonymous.

Keywords Climate change • Disasters • Policies • Gendered Practices • Aid relief

Introduction

Disasters of all kinds are increasing in frequency and the extent of damage caused. Aid giving has become big business. Although the size of the sector depends on definitions, what is covered and who collects the data, estimates suggest that \$128.7 billion was distributed in foreign aid by OECD countries in 2010 of which \$12

L. Dominelli (✉)
School of Applied Social Sciences, Durham University, Elvet Riverside 2, Durham DH1 3JT, The UK
e-mail: Lena.Dominelli@durham.ac.uk

M. Alston and K. Whittenbury (eds.), *Research, Action and Policy: Addressing the Gendered Impacts of Climate Change*, DOI 10.1007/978-94-007-5518-5_6,

billion was given as humanitarian aid for 'natural disasters' (One World 2011). Adding other donor sources, some estimates put humanitarian aid as high as $60 billion distributed by half a million relief workers annually. Also, definitions of disasters are becoming more contentious, as is attributing responsibility for causing them. For example, 'natural' disasters can be caused by climate change, e.g., the flooding associated with melting ice fields raising sea levels that are creating problems for: animal species such as polar bears and other marine life; indigenous peoples living around the Arctic Circle who are losing their livelihoods; and small island nations that are threatened with sinking into the sea (McGranahan et al. 2007). But climate change is primarily caused by human processes of industrialization and fossil fuel consumption. Thus, it is difficult to separate 'natural' disasters from (hu)man-made ones, including the contentious one of climate change, as Stern (2006) suggested.

In this article, I take the view that all disasters are interconnected, and that the deleterious nature of their impact is exacerbated for marginalized peoples, including indigenous people, children, women, older people, disabled people and those living in poverty. However, this consideration is not the main focus of this chapter. Instead, I consider gendered relations within climate change and disaster debates more generally to highlight the neglect of women's experiences during such calamities and make suggestions for its rectification in policy debates and practice. In doing this, I consider the experiences of women as recipients of aid and as relief workers to reveal that gendered relations silence the specific experiences of women. I draw largely on an Economic and Social Sciences Research Council (ESRC) funded project on humanitarian aid delivered during the 2004 Indian Ocean Tsunami in one of the worst affected countries in South East Asia, Sri Lanka (1), but keep the locations and identity of those involved anonymous.

Background

The tsunami of 26 December 2004 devastated lives, livelihoods and the physical environment in 12 countries bordering the Indian Ocean, causing 300,000 deaths. Sri Lanka had 40,000 of these (http://www.lankalibrary.com/news.htm). Aid giving is a contentious activity which has been accused of benefiting donors more than recipients. According to Hancock (1991) and Hoogvelt (2007), this is because much aid is inappropriate for local conditions and local residents are rarely consulted about what they want. Nonetheless, gestures of goodwill and the outpouring of humanitarian aid giving during this calamity were unprecedented. People living overseas were motivated to give in a way not seen since, including in the devastation of the 2010 floods in Pakistan and Australia; or earthquakes in Haiti, Chile, Christchurch or Japan since. Sri Lanka received a significant amount of this relief to enable it to recover from one of the most devastating natural disasters to strike in recent memory.

Hoogvelt (2007) argues that the neglect of local people as active citizens who can decide for themselves what they want in aid-giving processes makes humanitarian aid part of the problem rather than the solution. This, she suggests, is especially relevant in those parts of the globe where relief workers become embedded in the military physically as they did in Iraq or metaphorically by meeting people's basic needs and enabling people to survive conflicts that continue unchecked around them. In Sri Lanka, the armed struggle between the Sri Lankan government and the Tamil Tigers (LTTE) had been ongoing for several decades when the tsunami struck and the government restricted access to these areas for safety reasons. The ESRC research project did not cover LTTE controlled areas. Gender relations provide another salient issue for consideration. Researchers, including, Pittaway et al. (2007) revealed that women's concerns were often ignored and their needs poorly addressed in responding to the 2004 tsunami.

A range of overseas donors, civil society organisations, university personnel and relief workers arrived in Sri Lanka in significant numbers to provide funds and personnel to assist in the recovery process. Most overseas organisations left within 6 months. During their stay, they had considerable freedom to choose the sites that received their attention. Given their numbers, it is not surprising that some of these followed top-down forms of intervention that indicated that they knew best, while others endeavoured to be guided by the principles of empowering practices to place local people in control of developments. Two organisations forged links with local people soon after the tsunami and have continued to support them in a various ways since. To preserve anonymity, I describe one based in a university in the context of internationalisation (Callan 1998) as the institutional model. The other in a voluntary organisation provided the professional practice model. Both models sought to respond to local needs as expressed by local people in specific communities (Dominelli 2012a) adversely affected by the tsunami. Both models linked up with Sri Lankan universities and institutes interested in empowering local people. One also addressed the need to develop professional social work in the country.

Humanitarian Aid

The giving of humanitarian aid during disaster situations occurs in distinct phases. These are traditionally considered as preparedness, response, recovery and mitigation (Neal 1997). In this research, I identify them as follows:

- *Immediate aftermath*. During this phase, relief workers focus on providing survivors with food, water, medicines and shelter. There may be a glut of external agencies involved at this point, but most leave fairly quickly. Those involved in searching for people tend to be amongst this group.
- *Short to medium term assistance or recovery phase*. During this stage, financial resources, medical help, housing and psychosocial assistance are provided and a return to previous life routines is begun.

- *Long-term reconstruction and infrastructural development phase.* This focuses on permanent resettlement or return and involves rebuilding damaged infrastructures and physical environment.
- *Capacity building in institutions and people.* This promotes preventative measures that can mitigate risk in future and enhance preparedness in the event of a disaster.

The transitions and 'fit' between these phases were not usually thought out and poorly orchestrated. Women are absent as active players in all these stages and often miss out on aid entitlements unless they find people to advocate on their behalf or form groups to undertake such action themselves. Donors and recipients alike collude in this *gender silence*.

Humanitarian aid giving is a complex activity with a variety of models and ways of involving recipients in such interventions. I have classified these as falling into three major types:

- *Donor driven.* Those covered by this category are led mainly by external organisations that control the funds and resources that are brought into an area.
- *Recipient driven.* These forms draw primarily on local self-help groups and may suffer from a shortage of resources. They have the advantage of knowing the history, culture and language of an area.
- *Partnership between donors and recipients.* These involve a combination of local and external players, working in a range of partnerships varying from top-down power relations to egalitarian relationships that aim to empower local residents and respond to their needs as they see them. In the egalitarian version, solutions to problems tend to be coproduced by the overseas workers and local people working together.

Research Questions, Methods and Approach

This research project explored a number of questions about the delivery of humanitarian aid to obtain the views of local residents regarding their experiences of being recipients of aid. These questions were:

- Was the help offered what the people in Sri Lanka needed or wanted?
- Were some models of humanitarian aid more empowering than others? If yes, which ones? Why?
- Was the experience of receiving aid differentiated along local class, gender and ethnic relations impact on the aid received? If yes, how?
- Did the civil war impact on the receipt of aid? If yes, how?
- Can cross-country research be conducted in ways that enable local researchers to feel empowered and fully involved in research processes, especially when there are substantial differences and hurdles to overcome including those of culture, language, and access to resources when research is initiated by external actors?

This research project was informed by an ethnographic approach to ensure that the voices of Sri Lankan recipients of aid were heard. While the research team based in the UK spent time in Sri Lanka, it relied primarily on local Sri Lankan researchers to engage with villagers in the chosen research sites, getting to know the specifics of the local context, local residents, local governance structures and become accepted by them before the data collection stage of the research began. Involving local researchers as equal partners in research activities was part of the commitment to empowering practices in this study. Of the three local researchers, one was a woman. Their research was supported by one Sri Lankan consultant based in a Sri Lankan university. The Principal Investigator, a Co-Investigator and a Research Fellow were based in a university in the UK; and one consultant was based at a university in Eastern Europe. The research began in January 2009, several years after the tsunami. The passage of time before the commencement of the research enabled Sri Lankans in the affected areas to reflect upon the aid they had received in the immediate aftermath of the tsunami and subsequently. A variety of methods used to collect and triangulate the data. These included:

- Participant observation and conversations;
- Interviews of villagers, including residents of all ages, village elders, and officials (118 interviews in 11 villages);
- Interviews of staff and students who provided aid through the two models that were studied in depth (30 students and 10 staff from the institutional model; 10 students and 4 staff from the professional model);
- Mapping exercise of NGOs delivering aid;
- On-line survey of donors; and
- 4 Focus groups.

The interviews in Sri Lanka were conducted in the local languages by the local researchers and transcribed by them. They were then translated into English. Interviews in the other two countries (the UK and an Eastern European country) were undertaken in English and transcribed. The transcripts were cleaned, coded and entered into NVIVO software because this makes it easy to undertake analyses around specific questions. NVIVO was used to identify data relevant to the theme of gender relations. For the purposes of this paper, the findings were analysed according to this theme, with specific emphasis on the positioning of women within them. This revealed certain patterns of social exclusion generally, and of social relations within the family that positioned women as carers and restricted to undertaking specified roles within the domestic economy. These are described below, using the words of the interviewees to enable them to tell their own stories. This paper is based on interviews that had been transcribed and analysed at the end of the Project's second year.

The Project also had an Advisory Board consisting of representatives from practice, research and policy-making communities from four countries including the UK and Sri Lanka. Its role was to advise the researchers and comment on the coding schedules and findings. Some were also involved in dissemination events, including presenting papers at conferences examining disaster interventions.

Findings

In disaster situations, life is messy, complicated and does not always proceed according to plan. There was no one message coming from those who received aid. Key concerns that stood out in the analysis were: better planning to anticipate and mitigate some of the adverse outcomes arising from aid distribution; and giving women a voice and place in the proceedings. Doing this was expected to prepare people better for future potential disasters.

There were some critiques about how the overall distribution of aid was conducted. These respondents highlighted failings on the part of government, but wanted these addressed to prevent their reoccurrence in future and secure justice for some people who had not yet received compensation for their losses. These included complaints that some people received more than their share of aid and were better off after the disaster than before it. They claimed that people with connections in the right places got help more speedily and more in keeping with their needs. At the same time, some people did not get the assistance they needed, e.g., single men, women who had lost husbands and some fishing families. One villager's criticism focused on donors' failure to plan the distribution of aid, thus exacerbating existing inequalities and creating new ones by saying:

> The donors [should] closely supervise the distribution without just handing over to any person who comes forward. Otherwise, the real persons who should get support will not, [so] only the political friends and those who can spend a little money will benefit.

Women were amongst those who did not get their entitlements because they were excluded from aid distribution processes. And, unless asked specifically about women, their absence was not commented upon. Gender silence prevailed because little is said about women's roles or needs, thus enabling their rights to benefits to be ignored. Because little is said explicitly about women's contributions, they are the silent actors. However, the remainder of the paper reveals that women played key roles in keeping the family together and ensuring that its needs were met, even if theirs were not.

Speaking of the donors, some villagers defined their experiences of the distribution of aid as a mixture of supportive and unsupportive interventions:

> Some people were very good they treated all of us alike. Some people who came only helped their friends and political parties. Those people got more than what they had; others did not get anything. Sometimes their names were also struck from the list as they did not have money or did not go behind them.

Aid giving based on universality of treatment, i.e., all the same, produced unequal outcomes, even if much was positive. Silences about who was missing out exposed the lack of specificity about which groups were not covered and does not unpack differentiated gender relations. Yet, local people knew who they were and would comment about these if asked explicitly.

Internationalisation, or the interaction between the local and the global (and vice-versa) in the aid-giving processes, proceeded as exchanges between individual players rather than as planned strategy. These exchanges were not always experienced as reciprocal. In contrast, the institutional and the professional models were deemed valuable and empowering. These two models offered reciprocity, as is evident in the following comments by one villager:

> Actually [organization within the institutional model] brought them to the village, most of the villagers volunteered with their labour. [The] . . . students did all the nice painting on the walls. We taught them our dances as return. We made lots of handicraft items together. We taught them how to make them.

These comments also reveal the resilience that occurs within reciprocal exchanges. An extensive account of villagers' reactions is not covered by this paper, as it will be written about separately. This brief account is presented largely to give the globalised context (Lyons et al. 2006) within which these two models operated. One villager sums up their overall verdict on disaster interventions in general as follows:

> We learnt that if you want to survive you have to work hard. You have to have the money and right connections. There were good people and bad people. Some people liked to help while others looked at their benefit only.

As many external actors descended upon communities that needed disaster assistance, it can be difficult for local residents to distinguish between the different ones that come. The extent to which they remember specific agencies or people depends on the degree to which the external actors engage with the local people. This is what one villager said about those following the professional model:

> I can remember some of the white students [from Eastern Europe] who visited us . . . some months later . . . They came with students from [a Sri Lankan University]. There were also white teachers with them . . . Our young people were very happy to walk with them and visit the homes. They stayed two weeks, in a hotel in [local town] . . . Every morning during their stay they came to the village and visited the homes of the people . . . Some of them helped the poor people . . . even giving the people money to repair damaged houses . . . They taught our children English . . . told our young people about their customs. I can remember one of our boys had a birthday and they had a small party for him . . . they gave him flowers. They played on the beach with the children, our children liked to walk with them. They learned to talk a few words of English.

While remembering the specifics of particular groups of helpers can be difficult, villagers' experiences of their interventions is clearer. Interestingly, in this comment, ethnicity is emphasised while gender is not, although the majority of these students were women.

Masked Gendered Relations in Donor-Recipient Encounters: A Conspiracy of Gender Silence

Donor-recipient encounters appear gender neutral. They come across as an unplanned conspiracy of gender silence because gendered relations are not considered important enough to be commented upon. Thus, they become the unremarked upon background that shapes interactions between people and normalises them. This study revealed that overseas aid provided by the institutional and the professional model in the research sites was generally welcomed. In commenting on the latter, one villager states:

> It was good experience we got when we met the [Eastern European] students. They played with us. They sang songs; they worked with us to help build houses. We went to the beach and played on the shore. We had parties. They gave us nice flowers which made us happy and it was a new thing for us to give flowers.

New ideas and practices are created through such exchanges for both donors and recipients, even if this is not explicitly planned or evaluated (Knight 1997). The villagers' comments also indicate that gender relations remain masked while they highlight the importance of reciprocated exchanges. Most of the students being referred to were women engaging in what was seen as women's work. These comments also indicate that human interactions and responding to the human being in his or her situation is an important part of disaster intervention processes.

Treating everyone the same is seen as egalitarian treatment that does not discriminate against others. However, achieving equality is more complicated than this. Similarity of treatment presupposes that equality already exists, when this is not the case. Villagers commented upon their treatment in ways that ignored or masked gender relations although it was clear that men's and women's experience of the tsunami was different and that women's position and choices were much more constrained than those of men. Gender relations were assumed and taken-for-granted. Life was the way it was.

Gendered Relations in Disaster Interventions: 'Life Must Go on as Usual'

Gender relations continue to be problematic across the world (CSW 2009) and the picture is no different in climate change and other disaster interventions. In 1998, Enarson and Morrow wrote the *Gendered Terrain of Disaster* in which they identified the neglect of gender issues in relief endeavours, although women were found amongst the ranks of survivors and aid workers. Their concerns were replicated in this study. A woman aid worker interviewed in 2010 said:

> I looked at the field of gender and disaster management I went to all the possible NGOs who had worked in these areas and I couldn't get any report from any of the organizations with regards to that... and... it was women who were affected. Women's husbands died... and the women didn't have any way of surviving.

That the impact of gender relations on women's lives is regularly ignored is a reflection of socially constructed realities that continue to be reproduced during disaster interventions. This point of view was put by another woman interviewee who said:

> What men do and what women do is socially constructed and linked to cultural expectations about what it is OK for men to do and what women must do.

Consequently, gendered relations in disaster situations remain largely invisible. Life is expected to 'go on as usual', and is key to the view that people want to go back to their position as it was before the disaster (Clarke and Nicholson 2010). 'Bouncing back' was deemed key to resilience amongst individuals and communities and involves resuming a normal life, without change in either social or domestic relationships. As women's roles here have been articulated within the context of their family – an extended family, caring for family members, running the household and related activities linked to daily life routines, it is not surprising that getting back to normal resumes traditional gendered relations.

These comments reflect the reality of gender issues being configured around everyday life practices that are linked to women's ordinary routines in caring for children, relatives and husbands, doing housework and cooking. A woman interviewee put it thus:

> If men's economic activities were hampered they are excused by society.... Women are not excused by the society. They are supposed to do whatever... deal with the problems. They... take care of the household and the children.

With these words, this woman articulates how a gendered division of labour is alive and well in disaster situations.

Gender silences and patriarchal relations are often replicated through donor-led relationships. Hence, women's work is not valued and their needs remain unaddressed. Gender relations are expressed as taken-for-granted assumptions about what constitutes appropriate behaviour for men and women. These are reproduced by both donors and recipients except when women are specifically targeted by a few NGOs that aimed to enable women to fulfil their gender roles, including caring for children and elders, and engaging in income generation activities. Such targets re-entrench women's caring responsibilities through social policies and practices predicated upon gendered assumptions underpinning the division of labour, especially in the home. Sometimes, overseas NGOs and colonialists bring patriarchal ideas with them and introduce these into matrilineal societies where women are highly regarded (Stege et al. 2008).

Identities Are Expressed as Gendered Relations That Draw Upon Prescribed Roles and Existing Stereotypes

Identity discourses in disaster literature generally treat identity as homogeneous and unproblematic, leaving it ungendered, unclassed, unraced. Interviewees in this study problematised identity in ways that showed that gender, ethnicity and age are relevant in the social relationships that are formed in and through disaster interventions. These are usually traditional forms that reinforce patriarchal and racialised relationships because women are configured around their caring roles in the domestic arena.

Women's Roles

The roles that women and young girls can play in Sri Lankan society are shaped by expectations concerning their domestic responsibilities and cultural constraints upon their behaviour. However, women villagers did not express overt concerns about the limitations on their actions. These became apparent in what they did and did not do. The hard life that women faced is directly linked to men's economic dominance, as indicated in the comment below:

> In fishing villages women are in a lower position because men are very dominant. . . . In the south, there have been these traditional beliefs where women are not supposed to be on a fishing boat. It will get polluted. They are not supposed to touch anything, so women were completely excluded . . . .the domination that was built among men created many difficulties for women because alcoholism was very high. Men coming home from fishing in the morning. They drink and sleep. That's all they do. Wife battering is very common.

Driven by the needs of their families, women endure these situations to ensure that the family obtains what it needs. Although women may reject some of the roles ascribed to them, their resilience, in tough situations shines through. They will sacrifice personal autonomy and much else to keep their children safe and cared for, as the extract below indicates:

> She didn't want to get married. She didn't want to take a job in a garment factory which she has to because her husband drinks too much . . . and she's got a small child . . . and she's never there because she is always working in this garment factory and she wanted to get an education, she wanted to teach.

Women rarely expressed views critical of their position, or the hardships they experienced. They simply got on with life and came across as remarkably strong and resilient despite the demands that hardship and traumatic situations brought their way. Nonetheless, some women aid workers observed these restrictions and commented on them:

> The girls don't play cricket on their own without female chaperones. They can't, they won't. So essentially what we're doing is cementing the gender divide that already exists.

In such contexts, gendered relations impact extensively upon women's existence. Some gendered cultural constraints can endanger women's lives. One aid worker claimed that:

> Women and children were killed for social cultural reasons. I came across situations where women were almost saved and there was . . . a mother who was telling me that this girl was drowning and somebody gave her a hand to reach her. She was almost lifted and when she realized that she didn't have any clothes on her body . . . she let the hand go and she let herself drown . . . Her feeling of shame was much stronger than the feeling of death.

This account affirms the view that gendered cultural barriers are extremely powerful and difficult to change. Death is preferred over dishonour, which of course has implications not only for the woman or girl concerned, but also for her family.

Gendered cultural constraints were also evident in less life-threatening situations. As another aid worker commented:

> Women were refusing to enjoy their life. . . . Women were refusing to laugh even. They were feeling like how can we laugh because our people died? Our children died. This happened to us. We are not supposed to laugh.

Gendered cultural expectations are experienced as part of taken for granted everyday life routines and intractable. This was also evident where women were denied aid because they lacked husbands, fathers or brothers to look after their interests. As a result, many women did not receive aid to which they were entitled, a finding also made by Pittaway et al. (2007).

Masculinity

Masculinity and how men and boys express their masculinities is a significant part of gendered relations. In the research sites for this project, masculinity was not theorized as such. Popular conceptions of masculinity described men's and women's places in society, framed evaluations of young men's responses to the tsunami and defined their responsibilities towards their communities. Fishing is a male-dominated economic activity in Sri Lanka and expressions of masculinity were strongly tied into that. The men did the fishing. Losing their fishing boats was destructive of masculinity, and denied men the ability to act as providers within their household. As one villager said:

> Our boys were very good, they helped others who came to clear the dirt, bury the bodies, and clean the houses. Some of the young people had lost their boats and fishing things and they were sad. Some bad things also happened. Some of the boys got to drink arrack and began to fight. Some boys became lazy.

Whilst these comments reveal some villagers' concern about young men losing their way, they also indicated how individual responses to disasters and trauma vary considerably, and not always in socially approved ways. Some individuals are resilient and respond positively; others behave in ways that are destructive of communal harmony.

Racism and Ethnic Relations

Racism was not often discussed by interviewees. However, one overseas worker commented upon the unequal treatment of persons with minority ethnic origins:

> I personally feel that we have a form of preferential prejudice or . . . academic apartheid which we practice. So, for example, a visiting dignitary from the United States is likely to be accorded first class status as a visitor and treated in a certain way . . . but I'm not sure that . . . a professor of similar standing from a poorer country would . . . be treated in quite the same way. So I went from the very start wanting to make sure that our Sri Lankan counterparts did not feel that they were going to be treated as somewhat lower down the

> scale compared with someone from New York or Los Angeles... [who] gets more respect and attention than someone who comes from a poorer country perhaps because they are seen as being more relevant or within the academic structure [or that] they might be a partnership that's really worth growing and developing. A visitor from the United States would be put up in a very expensive first class hotel whereas a visitor from a poorer country might be asked to stay in a perfectly comfortable, perhaps more humble setting, like a simple college.

These comments raise the issue of status when allocating accommodation in aid endeavours and highlight how this and other factors including power relations, cultural differences and status symbols in both external and internal locations play out in humanitarian aid exchanges to produce, as this interviewee says, a form of academic racism. Forms of inequality seem difficult to overcome even when collaborators set out to recognise the contributions of all equally.

Dangers for Women Aid Workers

The literature on disaster interventions rarely focuses on the specific dangers that women aid workers face as women. In this research, women aid workers raised issues about their safety as women aid workers and revealed how being from a different ethnic (white) group worked to their disadvantage, as they lacked the protective factor that local ethnicity provided for Sri Lankan women. This resulted in sexual harassment and sexualisation of women simply because they were foreign nationals. One woman aid worker talked of her experience of being sexualized as follows:

> If you're on a bus a lot of times... they would try and touch you and put their hands up our skirts and down your tops... they would stare at us... and made you feel really uncomfortable.... We wanted to come back to [home country] just for that, just so the men weren't perving on us.

The sexual abuse of women aid workers is seldom referred to in the disaster intervention literature. There is a conspiracy of silence that is shared by the victim-survivors. This situation is not unfamiliar to women surviving violence in general (Hestor and Westmarland 2007), and when asked what they did to protect themselves from unwanted attention, the reply was: 'We slapped their hand and discouraged them... but we really didn't say anything'.

These women thought about how to address sexual harassment effectively. Ironically, when asked what they would like to see the donor agency do to protect them better, the women aid workers drew on traditional gender relations to find the solution. As one of them said:

> From a safety point of view, there were only two boys to share amongst thirteen girls... It was a remarkable difference to be walking down the street in a group of girls and to be walking with one of the boys because of the attention that you attracted as a white girl in Sri Lanka.

These women thought that the donors should attract more men into their humanitarian aid efforts and felt that a shift of emphasis in activities might get more men to join their ranks. As one woman aid worker put it:

> We had a lot of girl-centric skills, maybe more of the active roles, like sports, stuff that might have interested the guys.

The formation of intimate relationships between overseas aid workers and local people was not considered appropriate or advisable. Nonetheless, these did occur, sometimes with disastrous consequences for the women concerned. Below is one woman aid worker's account of what happened to one of her colleagues:

> [She] had a relationship with a Sri Lankan boy ... and had a real problem ... he hit her on the face ... They ruined the reputation of our project ... They ruined the reputation of foreigners because Sri Lankans already have got the wrong idea about foreign girls.

Such developments were considered serious because foreign women aid workers become further stereotyped and more endangered as a result. But there is silence around this issue.

Rescuing Children

Although children are considered a priority in disaster situations, they are seldom consulted or considered as a source of strength in disaster interventions, whether of a preventative, mitigative or adaptative nature. Children have strengths that should not be discounted. They like to contribute to rebuilding their homes and communities. They are also vulnerable during disasters and after. Children can be subjected to physical and sexual abuse or be trafficked by sexual predators (WHO 2004). Protecting children and rescuing them from harm are dominant themes in disaster discourses, with education playing a key role in such scenarios. Sending children to school as soon as possible is essential to settling them into a recognised routine (Seballos et al. 2011). One adult explained that 'They need 'protection', i.e., being rescued from harm'. Another adult interviewee commented that children relied on their mothers to cope with disaster:

> Children suffered a lot, they not only lost belongings but some of them lost one or both parents ... Children were caught by the waters and injured ... Some of them could not sleep at night ... They were scared to leave their mothers ... Some of the people who came [Red Cross/World Vision/FORUT/Christian Children's Fund] did some programs for the children ... They made them play games, made them draw, sing and dance ... Some organizations took the children to the beach. Many organizations and individuals gave the children clothes, school equipment, milk, foods, and other useful things ... Some school children even gave the children note books with their notes.

Children need specific forms of assistance. Schooling is crucial for them and essential in re-establishing normal routines that assist in their recovery processes.

Family support, especially that provided by their mothers is also crucial, and is one reason why social workers in disaster situations emphasise reuniting families as quickly as possible (Seballos et al. 2011).

Children, like women and other victim-survivors of disasters, are seen as a unitary, homogeneous category, instead of a differentiated group whose experiences of disasters differ according to age, gender, ethnicity, ability and class or caste which also shapes their experience of and reaction to both the disaster and helping processes. Dependency is important in configuring responses to children and can be a source of adults assuming power over children and getting them to do their bidding by withholding acceptance, approval and resources (Dominelli 2012b). One parent expressed their pleasure in seeing their children grow and adapt after the tsunami, including changing their relationship with the world in the following terms:

> I feel that my daughter is better now, all her improvements are from the village pre-school. She participated in two pre-school concerts, she is not shy, she is better in English and she sings well, so all these came through the village pre-school. Now we have got a new building and more kids . . . They get food from the pre-school . . . so, my opinion is that this pre-school is improving, we have to protect it and we have to send our kids to this one.

This parent worries about the future for this daughter and argues that benefits have to be protected and extended, even institutionalised. Gender appears in relation to the girl child as someone to be cared for, while the parent (the mother in this case) does the caring. However, the focus groups revealed that many children were looking after themselves, i.e., they were exercising agency and trying to recover from the disaster as best they could. Older children looked after the younger ones, particularly if parents were missing, until alternative arrangements could be made.

Some parents felt that the pre-schools built with donor funds were less well-resourced than those in nearby towns. One aid worker commented:

> They [the villagers] think the pre-schools in the town are better. Teachers there are so qualified and they teach better and the pre-schools are fully equipped.

The teachers in both locations are women. But their gender is not considered relevant. Villagers also claimed that if these gains were not protected, they could be lost. They raised the issue of external donors providing long-term planning and development when providing assistance.

External Actors Reflect Upon Their Own Interventions

External actors in disaster situations have the privilege of coming and going from a community that has suffered enormous loss during a catastrophic event whenever they wish. The same cannot be said of local residents whose lives are wrapped within a place and where recovering from severe destruction to and disruption of their lives is crucial. They lack the option of leaving. In reflecting upon the impact of their work in tsunami affected communities, one relief worker said:

The village community, they benefitted from the structural building there, they enjoyed having us there, and they probably learnt stuff about Western culture. We made friends in the village. I still think there is more that could be done….I think there are problems with the educational program there.

Reflections such as these indicate that some external actors can see a need for more thought to be given to capacity building in the educational system. Although the needs of children are important and meeting these remains crucial in all interventions involving aid-giving, women's needs continue to ignored. Women's needs as women must become articulated.

Who Benefits

Questions about who benefits from aid delivered during disaster interventions did not identify women as a specific group that either benefited or lost out from such assistance. One villager explained the situation as follows:

A few youths who had political support got boats even if they were not fishermen. So what they did was they sold these and spent the money on drinks. It was a waste. This happened because the people who were involved in the distribution did not select the right people. They took money and gave it to those people and their friends.

This person continues:

When the [Eastern European] students came there were many youths in the area who were very happy as they would take them to go visiting the houses, get them involved in whatever work they planned, like helping people to repair their houses. Play with them, sit with them and have their meals [with them].

This interviewee raises issues about nepotism and corruption, but moves on to focus on the positive aspects of aid workers from the professional model, although earlier comments revealed that not getting aid to those who need it can create considerable new difficulties for local people, e.g., establishing a drinking culture amongst young men.

Additionally, villagers suggested that relating to local people on their terms is significant to aid giving processes.

The students had helped the villages with donations, the students were not proud, they visited the simple homes of the poor people in the village and even have their food with the youth of the village whom they worked with..they visited our homes; sat with us. Village youth liked them… they went about with the youth, they even had their meals with them… they helped the poor families with gifts and money.

In other words, relief workers who engage with and become like the locals are more appreciated than those who do not. But, external actors are all considered a homogenous group.

Conclusions

The aim of humanitarian assistance is to improve the quality of life amongst local people and their communities in the immediate aftermath of a disaster and in long-term reconstruction. Unitary conceptualisations of identities and traditional gender relations exclude women from these processes. Humanitarian aid giving involves internationalising practices that are complex and involve many actors and agencies, often in contradictory relationships. Within this, observing local cultures, traditions and languages is crucial to the aid processes, but these may not help women because gendered social relations remain invisible and can be ignored.

The unrecognised contributions of women and the failure of aid giving processes to affirm women's entitlements and rights indicate that the values that humanitarian aid organisations espouse, especially those involving equality and empowerment, are difficult to realise in practice across all groups and settings. The silence around gendered relations makes listening to women and children important and involving them directly in aid distribution processes is long overdue. Capacity building for long-term sustainable development is essential and should be taken on board at the beginning of humanitarian aid endeavours. Women should be involved in these activities from the word 'go' and throughout, including in their evaluation.

The interviewees in this research identified the following general areas for improvement:

- More effective assessment of needs.
- Assessing the needs of men, women and children as gender and age specific.
- Criteria around entitlements should be inclusive, transparent and properly implemented.
- Transparency in aid distribution processes at all levels.
- More equitable systems of resource allocation and distribution.
- Long-term sustainability of initiatives considered from the beginning of the humanitarian aid process.
- Monitoring mechanisms to ensure that all targets have been achieved and those needing help have received it.
- Locally driven processes and power-sharing between external donors and local recipients.
- Humanitarian aid as a constantly evolving, adaptable and resilient process with women, men and children at its heart.

Making the differentiated experience of women, men and children explicit in disaster intervention processes is essential if all voices are to be heard. These ideas should be incorporated into debates, policies and practices around disasters including climate change. Their implementation can: prevent further exclusion; develop community resilience during disasters, after the recovery and reconstruction processes; and recognize and value women's contributions to these endeavours at all stages.

Acknowledgements Thanks to Dr Tom Vickers, the Research Associate for his assistance in extracting the quotes used in this article from the NVIVO system. Other acknowledgments include Joy Palmer-Cooper (CI); the local researchers – Sunil Shantha, Kanthi Perera; and consultants Subhangi Herath and Bogdan Lesnick.

References

Callan H (1998) Internationalisation in Europe. In: Scott P (ed) The globalisation of higher education. OUP, Buckingham

Clarke J, Nicholson J (2010) Resilience: bounce back from whatever life throws at you. Crimson Publishing, Richmond/Surrey

CSW (Commission on the Status of Women) (2009) 2009 Survey on the role of women in development. CSW, New York

Dominelli L (2012a) Globalisation and indigenisation: reconciling the irreconcilable in social work? In: Lyons K, Hokenstad T, Pawar M, Huegler N, Hall N (eds) The sage handbook of international social work, Sage Publications, pp. 39–55

Dominelli L (2012b) Green Social Work. Cambridge, Polity Press

Enarson E, Morrow B (1998) The gendered terrain of disaster: through women's eyes. Florida International University, Miami

Hancock G (1991) Lords of poverty: the power, prestige, and corruption of the international aid business. Atlantic Monthly, New York

Hester M, Westmarland N (2007) Domestic violence perpetrators. Crim Justice Matters 66:34–39

Hoogvelt A (2007) Globalisation and imperialism: wars and humanitarian intervention. In: Dominelli L (ed) Revitalising communities in a globalising world. Ashgate, Aldershot

Knight J (1997) Internationalisation of higher education: a conceptual framework. In: Knight J, De Wit H (eds) Internationalisation of higher education in Asia pacific countries. EAIE, Amsterdam

Lyons K, Manion K, Carlsen M (2006) International perspectives on social work: global conditions and local practice. Palgrave Macmillan, Basingstoke

McGranahan G, Balk D, Andersen B (2007) The rising tide: assessing the risks of climate change and human settlements in low elevation coastal zones. Environ Urban 19(1):17–37

Neal D (1997) Reconsidering the phases of disaster. Int J Mass Emerg Disasters 15(2):239–264

One World (2011) Humanitarian aid. http://uk.oneworld.net/guides/aid?gclid=CJ32vvW6uKwCFdQOfAodLWQJyw#Statistics

Pittaway E, Bartolomei L, Rees S (2007) Gendered dimensions of the 2004 Tsunami and a potential social work response in post-disaster situations in international social work, 50(3):307–319

Seballos F, Tanner T, Tarazona M, Gallegos J (2011) Children and disasters: understanding impact and enabling agency. IDS, Brighton

Stege K, Maetela R, Naupa A, Simo J (2008) Land and women: the matrilineal factors. Pacific Islands Forum Secretariat, Suva

Stern N (2006) Stern review of the economics of climate change. Cambridge University Press, Cambridge

WHO (World Health Organisation) (2004). Preventing violence: implementing the recommendations of the World report on violence and health. Geneva: WHO. Also on www.who.int/violence_injury_prevention/media/news/08_09_2004/en/. Accessed 11 Nov 2011.

Chapter 7
Gender, Development, and Rights-Based Approaches: Lessons for Climate Change Adaptation and Adaptive Social Protection

Beth Bee, Maureen Biermann, and Petra Tschakert

Abstract As adaptation policies increasingly aim to recognize the agency and autonomy of marginalized groups, the push for gender-sensitive climate change policy is also gaining voice and visibility. This shift is akin to the growing popularity of rights-based approaches to development, suggesting that lessons can be drawn from previous gender and development debates to inform adaptation policy and practices. In this chapter, we explore the nexus between gender, development, and right-based approaches in order to highlight the possibilities and pitfalls of such an approach to adaptation. We examine recent (adaptive) social protection programs to understand the potential of rights-based approaches to build capacity and enhance resilience under climate change. Finally, by emphasizing social responsibilities to and for others, we indicate the opening of spaces for a more inclusive social project for promoting adaptation that values differential skills, assets, expertise, and voices while acknowledging the limits of autonomous actors in adaptation.

Keywords Gender • Climate change adaptation • Rights-based approaches • Social protection • Justice

---

B. Bee, Ph.D. (✉)
Josef Korbel School of International Development, University of Denver, 2201 S Gaylord St., Denver, CO 80208, USA
e-mail: bbee@du.edu

M. Biermann, M.S • P. Tschakert, Ph.D.
Department of Geography, Pennsylvania State University, 302 Walker Building, University Park, PA 16802, USA
e-mail: biermann@psu.edu; petra@psu.edu

M. Alston and K. Whittenbury (eds.), *Research, Action and Policy: Addressing the Gendered Impacts of Climate Change*, DOI 10.1007/978-94-007-5518-5_7,

Introduction

In recent years, scholarly work and applied case studies that examine the impacts of global environmental change in local places have made a gradual shift from vulnerability assessments to an emphasis on strengthening adaptive capacities to reduce vulnerabilities. Adaptive capacity is understood as "the potential of individuals, communities, and societies to be actively involved in the processes of change, in order to minimize negative impacts and maximize any benefits from changes in the climate" (Pettengell 2010, p. 7). Creating less vulnerable, more resilient individuals, societies, and communities by enhancing the mechanisms that promote adaptive capacity is an important element in the adaptation literature (Folke et al. 2002; Smit and Pilifosova 2003; Adger et al. 2005; Smit and Wandel 2006; Berkes et al. 2007; Tschakert and Dietrich 2010). Rather than categorizing marginalized groups as passive victims, a focus on adaptive capacities recognizes the agency and autonomy of those most affected by climate variability and change. This shift in the adaptation literature also includes both acknowledgements of the gendered differentiated impacts of climate change, as well as concerted efforts to alter the climate change rhetoric that negates women's agency by constructing them as vulnerable victims (MacGregor 2009; Dankelman 2010; Arora-Jonsson 2011; Cuomo 2011; Resurreccion 2011). At the same time, a broader call to address gender in climate change policy has been taken up by several mainstream development organizations (e.g. the United Nations' Food and Agriculture Organization, United Nations Development Programme, Women's Environment and Development Organization and United Nations International Strategy for Disaster Reduction).

This trajectory toward recognizing the agency and autonomy of marginalized groups, including women, is akin to the growing popularity of rights-based approaches to development. Although it is not without its controversy, for many, rights-based approaches offer a means to secure structural change, address power inequities, and protect the poor (Hickey and Mitlin 2009). Based on lessons learned from gender and development debates, this paper explores the nexus between gender, development, and right-based approaches to highlight both the potentials and the pitfalls of a rights-centered lens to climate change adaptation. We examine recent adaptive social protection programs, as an integrated approach to promoting climate-resilient livelihoods and anchored by a right-based framework for addressing inequity and injustice around climate change.

The first section of this chapter reviews current scholarship that draws our attention to the problematic and essentialist ways that marginalized groups, particularly women, are constructed in climate change discourses; it also examines adaptation programs that address the underlying causes of vulnerabilities. We suggest that a rights framework is one potential means to simultaneously subvert these essentialist constructions and explicitly address unequal power relations that perpetuate and sustain inequalities that are often at the root of larger, endemic social problems of poverty and exclusion. The next section of this chapter therefore reviews the debates, particularly those grounded in feminist thinking, around rights-based approaches to development to deepen our understanding of the opportunities and drawbacks of applying a rights-based approach to climate change adaptation. Drawing on the lessons learned from these debates, we then examine how recent adaptive social protection programs, grounded in a rights-based framework, are one potential avenue for both recognizing and realizing the rights, agency, and autonomy of marginalized groups. We conclude by reflecting on the potential that adaptive social protection programs hold for correcting gender biases, building capacities, and enhancing livelihood resilience under climate change. By putting emphasis on social responsibilities to and for others, adaptive social protection programs can open spaces toward a more inclusive social project that values differential skills, assets, expertise, and voices while acknowledging the limits of autonomous actors in adaptation.

Climate Change Adaptation and Gender

Two decades of work on climate change vulnerability within a broader development discourse has made it increasingly obvious that climate change adaptation requires tackling the underlying causes of vulnerabilities and inequities. In order to be successful, fair, and sustainable, climate change adaptation policies and programs must necessarily engage with issues of gender. Lambrou and Piana (2006) eloquently demonstrate how gender relations determine social conditions that leave millions of women economically insecure, overborne with care giving responsibilities, lacking social power and political voice, and therefore with little capacity to adapt to any kind of stressor, shock, or change. In addition, they argue, gender biases in planning bluntly reinforce these inequities as men's voices are prioritized over women's in decision making, rendering women and women's concerns virtually invisible in the policy-making arena. Multiple scholars have similarly argued that power relations and differential access to resources are key to understanding gender differentiated vulnerability, exposure to risk, coping capacity, and the ability to recover (Masika 2002; Tandon 2007; Rossi and Lambrou 2008; Nelson and Stathers 2009).

Women's experiences with climate change adaptation and livelihood resilience under multiple stressors have been largely disregarded within the broader climate change literature. Despite growing attention to gendered knowledge systems, skills, resources, and specific coping and adaptation strategies that women have employed in the face of climatic stressors (e.g. Boyd 2002; Ahmed and Fajber 2009; Buechler 2009; Segnestam 2009; Petrie 2010; Alston 2011; Glazebrook 2011; Onta and Resurreccion 2011), the debate has remained, to a large extent, focused on women's vulnerability to harm, and especially of poor rural women in the Global South. An overemphasis on women's vulnerability, albeit politically and strategically useful, risks diverting attention from inequalities in decision-making.

While women certainly represent no homogenous group – neither as universally vulnerable in the Global South nor as intrinsically virtuous and environmentally savvy in the industrialized North, most recent literature on gender and climate change cautions against the continuous overemphasis of women's vulnerability in as far as it obscures power imbalances and denies women agency while legitimizing differences as unchangeable. For example, Arora-Jonsson (2011) argues that the counterpart to women's vulnerability, their virtuousness or their propensity for being more environmentally friendly or open to social change, is as reductionist as the victim discourse and equally deflects attention from power relations and inequalities reproduced in institutions and discourses on climate change. Furthermore, Resurreccion (2011) rejects the discourse of women as "chief victim-and-caretaker" (p. 3), arguing that it uncritically perpetuates the essentialist and oversimplified view propagated by the Women, Environment, and Development community (WED) in the 1980s and 1990s that highlighted women as natural constituencies for environmental projects against land degradation. MacGregor (2010) points to the fact that a focus on women's vulnerability to climate change serves to construct women, especially in the Global South, as "one dimensional objects" that populate the debate only as helpless, voiceless, and unable to cope without the help of large development organizations (p. 227), thereby exacerbating negative stereotypes. This specific construction further perpetuates an "us" and "them" attitude whereby climate change becomes a problem for "them", not "us" (MacGregor 2010).

Additionally, such disempowering discourses can be and have been used to legitimize the imposition of external, top-down interventions to 'effectively' tackle climate change that further marginalize and silence women's as well as male voices and their experiences in dealing with climatic extremes. Li (2007) demonstrates that similar constructions of the "other" have been used to justify imposed interventions throughout the history of colonialism and post-colonial development. This type of victim discourse has been continually debated by feminist scholars who contest such totalizing constructions of women, particularly of third world women of color (Moraga and Anzaldua 1984; Ong 1988; Mohanty et al. 1991).

In order to counterbalance the persistent women as "chief victim-and-caretaker" discourses (Resurreccion 2011, p. 3) efforts have been undertaken at the policy level to 'mainstream' gender into development policies as well as disaster risk reduction and climate change adaptation programs. Drawing upon gender mainstreaming in development, where it has long been the norm, gender-sensitive policies attempt to unravel gender-differentiated impacts of climate change and design appropriate prevention and response strategies. However, such attempts can result in an uncritical addition of women and gender-sensitive language to development projects that do not challenge the underlying structural causes of gender inequities (Woodford-Berger 2004; Walby 2005). Simultaneously, non-governmental development agencies and activist groups have attempted to re-imagine vulnerability by depicting women and groups typically categorized as vulnerable as "positive icons" and "speaking subjects" that reverse the colonial binary reproduction of the superior North and the inferior South. Manzo (2010), for instance, in an examination of major climate action campaigns, recommends de-emphasizing vulnerability by highlighting cultural adaptability and resilience and de-emphasizing difference by stressing connectedness. Despite these achievements, important work remains to be done that focuses explicitly on the analysis of inequalities and power relations between men and women and how they influence gendered access and control over resources and household and community decision-making processes in the context of climate change adaptation (Glazebrook 2011; Bee 2012). Furthermore, the gender myths that have been employed by development agencies that essentialize women's roles and characteristics and simplify human behavior must also be addressed (Cornwall et al. 2007).

One proposed means to address these myths and re-construct the relationship between adaptation and vulnerability is through an integrated approach that includes elements of rights-based approaches to development. Such approaches can be useful for addressing power relations that sustain and perpetuate the inequalities that are often at the root of the larger endemic social problems of poverty and exclusion. These are also the same relations of power that hinder fair and timely adaptation to climatic and other shocks and crises. An integrated approach, grounded in a rights framework, has the potential to address the structural causes of poverty and vulnerability, without casting vulnerable populations as passive victims (Symington 2002).

Consequently, we argue that rights-based approaches to climate change adaptation warrant further examination, as they may be an effective means to simultaneously promote social justice and address the underlying causes of vulnerabilities. Feminist scholars have created a significant body of academic research that scrutinizes the language of gender representation, equity, empowerment, and rights in development, and this work has been influential in development institutions. Now, as adaptation begins to intertwine with development priorities and practices, lessons can be drawn from feminist debates about both the possibilities and the limits of rights-based approaches to inform more just policy and practices in climate change adaptation.

Feminist Perspectives on Rights-Based Approaches to Development

Proponents of rights-based approaches to development argue that they bring people's actions and agency to the forefront, and that political processes within development become central to discussions and strategies (Moser and Norton 2001; Molyneux and Lazar 2003). The recognition of women's rights has been most salient in research and policy on women's reproductive rights, gender-based violence, political violence, and peace building (McIlwaine and Datta 2003). However, the positive strides taken within the rights-based discourse have not been without criticism. Principal among these critiques is a discomfort with the tendency to focus on negative rights (the right to be safe "from") instead of positive rights (the right "to") (McIlwaine and Datta 2003). Concerns also arise over a disconnect between the establishment of and the actual exercise of rights, and that the discourse of rights is often conceptualized in Western, liberal, and therefore individualistic terms which can be at odds with places and communities that value communal and group rights (Hernández Castillo 2002; Radcliffe 2002). Feminist theorists themselves are divided within the broader debates over universalism and multiculturalism as to which principals are the most effective for achieving equality (Molyneux and Razavi 2002). As a result, there are often tensions between the language of universal rights and the context-specific local practices. The manner in which a given concept of rights is deployed also impacts outcomes. Following an extensive analysis of the ways that mainstream development organizations deploy rights-based approaches to development, Cornwall and Nyamu-Musembi (2004) identify a typology of four ways a concept of 'rights' may be deployed: as a set of normative principals; as a set of instruments; as a component to be integrated into programming; and as the underlying justification for interventions aimed at strengthening institutions. Using only one type of deployment may limit the possibility for wider and deeper social transformation. On the other hand, incorporating all four tactics into rights-based practice can counteract such shortcomings (Cornwall and Nyamu-Musembi 2004).

Looking specifically at what 'rights' can contribute to struggles for gender justice, several feminist scholars explore the dilemmas of the politics of rights in a special issue of *Third World Quarterly* (Cornwall and Molyneux 2006, p. 1175). These articles raise questions about the legitimacy of a Western-inspired agenda of liberal rights and its fit or friction with culturally specific conceptualizations of rights in non-Western places and the efficacy of a focus on rights as a means to advance equality and social justice. In their introduction, Cornwall and Molyneux suggest that the term 'rights-based' offers development organizations a means to legitimize their efforts: "[it] provides moral authority and purpose, justifies intervention and gives succor to those concerned with enhancing the capacity of those who have been marginalized to claim rights and recognition" (2006, p. 1179). Drawing on the work of Robinson (2005), Cornwall and Molyneux also suggest that rights complement mainstream development efforts to promote good governance and participation. However, they emphatically argue that the notion of rights, as embraced by mainstream development organizations, needs to be unpacked. Only through a critical analysis can we begin to address questions such as: What kinds of rights have been promoted? Why are these rights promoted and not others? For whom are these rights promoted? Whose rights have been compromised in the process? Even as these potential shortfalls and limitations are acknowledged, many development scholars remain hopeful of the promises that rights-based development holds. What is clear, as Cornwall and Molyneux point out, is that the political and social contexts within which activists and NGOs are working shape the attitudes about and effectiveness of rights-based approaches. As a result, which rights matter and how best to advance them are less about theory than they are about the material and political constraints that advocates encounter in various places (ibid.). Therefore, the appeal of a rights-based framework is not that it provides easy answers that simplify decision-making, but that it presents new opportunities to navigate policy and decision-making waters in a way that constructively engages inequalities by opening up new spaces for conversation and contestation. Such spaces and opportunities are sorely needed in climate change adaptation research, practice, and policy.

We now turn to explore the potential of adaptive social protection, a concept and practice grounded in a rights framework, as a means to strengthen the capacity for women and other marginalized groups to adapt to climate change by laying claim to full participation in adaptation programs, especially those that promote livelihood resilience rather than just agronomic or infrastructural adjustments. We suggest that adaptive social protection may provide a useful approach for gender justice in adaptation policy and programs, which have tended to overlook women as active participants in household and community adaptation. As demonstrated by the way rights-based approaches were mobilized in development policy, applying such approaches to climate change has the potential to address questions of gender justice by prioritizing gender equity concerns in a way that frames gender equity as a central tenet of human rights, instead of as "in addition to" (Goonesekere 1998).

Pursuing Rights Through an Integrated Framework: Adaptive Social Protection Programs

Reflecting on similarities and differences between rights-based approaches to development and climate change adaptation is a prerequisite for designing practical spaces in which rights and needs of vulnerable and marginalized people can be identified and negotiated. In order to avoid confining rights debates to legal and theoretical arenas or gender mainstreaming into "quiet complacency" (Drinkwater 2009, p. 146), development organizations such as CARE International, Oxfam, and Practical Action have embarked on innovative trajectories to anchor the language of rights and the importance of gender in their adaptation programs. Placing rights and gender at the center of these projects not only pays specific attention to the inequitable distribution of resources, power, and repressive cultural norms and rules that hinder people's ability to adapt to climate change; it also stresses the need for diverse skill sets to anticipate, prepare for, and respond to stressors and shocks, enabling and supporting institutions, and social protection programs as back-up plans when response strategies fail (Oxfam 2009; Oxfam and UN Vietnam 2009; CARE 2010; Practical Action 2010; Nelson 2011). An integral part of these programs is the explicit effort to encourage vulnerable populations to make transformational changes to their livelihoods that allow them to actively pursue opportunities and minimize negative impacts from climate change (Pettengell 2010).

In practice, many of these rights and gender-centered approaches to adaptation draw on the concept of adaptive social protection. Adaptive social protection combines three overlapping approaches: social protection, disaster risk reduction, and climate change adaptation. Building on rights-based discourses, and the shift from needs to rights in development thinking (Hickey and Mitlin 2009), the integration of these three approaches opens a critical space for empowering the poor and the marginalized to exercise their voice, claim the right to protection, and shape processes of social transformation (Conway et al. 2002; Heltberg et al.

2009; Arnall et al. 2010). After a brief explanation of the two core approaches not yet discussed, social protection and disaster risk reduction, we examine how recent adaptive social protection programs employ a rights framework to rectify gender biases, build people's capacity to adapt to sometimes unpredictable change, and enhance livelihood resilience to climatic and other stressors.

Social protection describes all public and private initiatives that provide income or asset transfers to the poor, protect the vulnerable against livelihood risks, and enhance the social status and rights of the marginalized (Devereux and Sabates-Wheeler 2004). Unlike food subsidies in the 1960s–1970s and food-for-work programs in the 1980s, more recent social protection programs focus on 'predictable funding for predictable needs' that allow regular cash and asset transfers to the poorest and most vulnerable rather than ad hoc and crisis-dependent deliveries (Ellis et al. 2009). However, social protection has often been defined and employed through a narrowly specified safety-net function, based on economic efficiency that merely reinforces coping mechanisms (Devereux and Sabates-Wheeler 2004). Yet, while some argue that social protection is purely an instrument to satisfy needs, particularly food security and other Millennium Development Goals, others advocate for social protection from a rights-based perspective by stressing food security and as an inalienable human right. Such a "universal social minimum", as argued by Davies et al. (2009), is threatened under climate change and therefore needs to be secured and enhanced. A key feature of *adaptive* social protection is that it includes elements that seek to transform unequal social relations, thereby addressing the underlying causes of vulnerabilities.

Another core feature of adaptive social protection programs is the integration disaster risk reduction strategies. Disaster risk reduction, once focused on humanitarian relief and reconstruction after rapid onset natural disasters such as earthquakes, hurricanes, and tsunamis, has moved more toward the prevention and reduction of risk prior to harmful events. The attempt is to make individuals and communities less vulnerable to shocks, mainly through the buffering of livelihoods, for instance through micro-credit, crop insurance, and livelihood diversification. Such a conceptual shift to tackling underlying social differentiated vulnerabilities and long-term pro-active disaster preparedness is laudable; however, in reality, reducing impacts still prevails over reducing vulnerabilities (Schipper 2009). Another lingering concern is that DRR programs tend to focus on returning to normalcy, although normal conditions are often those that create and perpetuate risk, injustice, and vulnerability (O'Brien et al. 2008). Similarly, Ray-Bennett (2010) cautions that providing access to microcredit, as a DRR strategy for asset building, can hardly reverse historic social inequalities regarding gender, class, and caste that render certain groups intrinsically more vulnerable.

Despite certain shortcomings in the respective elements (social protection, DRR, and climate change adaptation), the adaptive social protection framework moves our collective attention to a dynamic notion – hence the term "adaptive" – of actions and interventions, triggered by both the variability and uncertainty of climate change, including extreme events, that challenge the way short-term rights in disaster management and food security and long-term, enduring rights in development are constructed and implemented. We argue that there are three main reasons that adaptive social protection is a conceptually intriguing and practically relevant concept for fostering gender justice under climate change adaptation: its focus on dynamic and multi-dimensional vulnerabilities, its incorporation of a rights-based framework that stresses equity and justice, and its attention to transformation as a means of tackling the underlying structural causes of vulnerability. We provide examples for each of these below, drawing mainly from a meta-analysis for South Asia (Arnall et al. 2010).

First, adaptive social protection has shifted the focus from the chronically poor and economic safety nets in the social protection framework, to dynamic and multi-dimensional vulnerabilities that emerge or are exacerbated as a consequence of slow and rapid onset climatic events. This allows for protection and other immediate social responses so far overlooked in adaptation programs. Some climate events may not result in full-blown disasters such as floods or droughts but nevertheless can push disadvantaged individuals and groups beyond a critical threshold of resilience, for instance through altered cropping patterns (Tschakert et al. 2010). A community-based adaptation project run by the Himalayan Climate Centre in Nepal has the two-fold objective of increasing communication between National Meteorological Service of Nepal and local community institutions, and establishing a disaster insurance scheme (Arnall et al. 2010). The outcome of this linkage is to reduce disaster impacts and to promote adaptive local responses to stresses and shocks. While the overall goals of this project do not rely on or address gender differences, it lays the groundwork for flexible and adaptive social responses that push the boundaries of how vulnerabilities are understood and how adaptation takes place.

Second, through its rights-based perspective to vulnerability reduction, adaptive social protection addresses equity and justice dimensions in climate change adaptation and resilience, thus promoting social inclusion of often marginalized populations – women, elderly, indigenous, disabled, class, caste – to counteract oppressive socio-cultural, political, and institutional paradigms and practices. For instance, the National Rural Employment Guarantee Scheme (NREGS) in India, embedded in a rights-based framework, guarantees 100 days of work per rural household per year upon demand, at a minimum agricultural wage, paid in cash (Arnall et al. 2010). Adaptation research has shown that income (as well as cash transfers) is critical to allow for consumption smoothing while preventing the erosion of essential productive assets, which typically leads to poverty traps (Heltberg et al. 2009). At the same time, although the evaluations of cash transfer programs are inconclusive, it is generally accepted that the benefits of such programs depend a great deal on the socio-cultural context, especially gender relations and intra-household decision-making power (Devereux 2006)

Third, by promoting transformation, adaptive social protection programs can directly tackle structural causes of vulnerability that constrain people's adaptive responses under climate change. This may involve securing access to land for landless farmers, land transfers, and microenterprise development. The Churia Livelihood Improvement Programme in India, for example, funded by CARE International, works specifically with poor, vulnerable, and socially excluded women to address structural inequalities that inhibit adaptation (Arnall et al. 2010). Several Practical Action projects in Bangladesh are likewise aiming to transform local livelihoods instead of boosting short-term coping strategies, and are working to enable community participation in planning processes and institutions at multiple scales (Arnall et al. 2010; Practical Action 2010). This is critical, as coping often reinforces rather than overthrows encrusted layers of inequality that constrain successful adaptation. Similarly, the Adaptive Agriculture Programme implemented by the National Centre for Scientific Research (CNRS) in Bangladesh takes a transformative approach that aims to secure access to fallow lands for landless farmers (Arnall et al. 2010). A gender aware approach could also work to address the gender inequities that exist in both access to and control over land. Stribu (2010) further argues that strategies which work to challenge inequitable patterns of resource allocations, resource management and access to resources through distributive justice measures also provide long-term livelihood and socio-economic security.

These examples from adaptive social protection programs demonstrate the importance of including opportunities for individuals and communities to articulate their needs and rights in adaptation projects, in order to take into account underlying and often gendered causes of differential vulnerabilities. The explicit focus on asset protection and promotion mirrors similar frameworks (e.g. Oxfam and UN Vietnam 2009) that address gendered entitlements and capabilities, gendered norms and division of labor, and gendered perceptions of risk and interpretation of climate and climatic changes. Adaptive social protection programs have the potential to strengthen claims of marginalized groups that go to the heart of persistent and structural inequalities that repudiate gendered rights in the first place. Such an approach is in stark contrast to scores of adaptation projects that focus on technical or environmental solution to climate change, many of which may very well exacerbate existing gender inequalities (e.g. Carr 2008). Yet, we are aware that gender roles in climate change adaptation are fluid, negotiated, and even contested; what matters most for gender justice under climate change is how women and men can equally access and control decision-making processes and challenge discriminatory institutional practices.

Conclusions

This chapter has demonstrated why approaches to building people's capacities to adapt that are grounded in a rights framework present a promising, albeit not easy way for tackling persistent inequalities in power relations, access to resources, and decision-making processes that constrain effective, timely, and just responses to the negative impacts of climate change. Adaptive social protection appears to be a practical way for accomplishing this by emphasizing the rights and agencies of groups who are vulnerable to climatic and multiple other economic, political, ecological, cultural, and institutional stresses. Yet, as advocated by O'Brien et al.

(2004) and Nightingale (2009), among others, it is equally important to scrutinize the larger political economy and the use of discourses that cause and exacerbate vulnerability and inequalities, especially through the lens of gender, class, and caste. We must further examine what particular mechanisms and power structures are at play that result in discriminatory adaptation strategies, what alternatives are available, and what preferences both men and women identify for themselves that strengthen their own claims.

A rights perspective, at its core, addresses power inequalities, identifies ways to secure structural change, and attempts to protect the poor (Hickey and Mitlin 2009). Yet, the burden should not only fall on those vulnerable, marginalized, and poor who we – the distant advocates – expect to claim these rights. On the flip side of any claim to rights is an associated argument for the fulfillment of responsibilities. Anand and Gasper (2007) point out that even legally recognized claims to rights do not necessarily translate to those rights being satisfied. To be successful, a rights-based approach to adaptation, such as adaptive social protection, must therefore simultaneously address questions of rights and the associated responsibilities. In order to include this emphasis on social responsibilities to and for others, we indicate the re-opening of spaces for a more inclusive social project that values differential skills, assets, expertise, and voices while acknowledging the limits of regulating actors. Although the above discussion of adaptive social protection suggests some practical examples of how this may be done within the context of adaptation projects in the Global South, we strongly encourage attention to the gendered roles of green consumerism and energy use in the North and related climate and justice debates (e.g. Terry 2009; MacGregor 2010; Cuomo 2011). We also highlight claims from rights-based approaches to development that advocate for social responsibility to and for others, including the close and distant vulnerable and poor, as an inclusive project of social justice (Hickey and Mitlin 2009).

More critical and engaged work needs to be done to creatively imagine rights to access resources and rights for fair political representation and participation to pursue adaptation strategies that neither heighten existing inequalities nor enhance women's responsibilities without reward. We suggest that adaptive social protection represents one example of a rights-based approach that leaves opportunity to address questions of social and gender justice. At the same time, we underscore the need to direct attention in a concerted effort to the larger structural drivers of localized vulnerability and marginalization, as well as the fluid and often contested roles that men and women play in preparing for and responding to the challenges of climate change.

References

Adger WN, Arnell NW, Tompkins EL (2005) Successful adaptation to climate change across scales. Glob Environ Chang 15:77–86

Ahmed S, Fajber E (2009) Engendering adaptation to climate variability in Gujarat, India. Gend Dev 17(1):33–50

Alston M (2011) Gender and climate change in Australia. J Sociol 47(1):53–70
Anand PB, Gasper D (2007) Guest editorial: conceptual framework and overview. Special issue on human security, well-being and sustainability: rights, responsibilities and priorities. J Int Dev 19:449–456
Arnall A, Oswald K, Davies M, Mitchell T, Coirolo C (2010) Adaptive social protection: mapping the evidence and policy context in the agriculture sector in South Asia. IDS working paper 345, Institute of Development Studies, Brighton. http://www.ids.ac.uk/files/dmfile/Wp345.pdf. Accessed 30 Jan 2012
Arora-Jonsson S (2011) Virtue and vulnerability: discourses on women, gender and climate change. Glob Environ Chang 21(2):744–751
Bee B (2012) Who reaps what is sown? A feminist inquiry into climate change adaptation in two Mexican ejidos. ACME Int E J Crit Geograph (in press)
Berkes F, Armitage DR, Doubleday N (2007) Adaptive co-management, collaboration, learning, and multi-level governance. UBC Press, Vancouver
Boyd E (2002) The Noel Kempff project in Bolivia: gender, power, and decision-making in climate mitigation. Gend Dev 10(2):70–77
Buechler S (2009) Gender, water, and climate change in Sonora, Mexico: implications for policies and programs on agricultural income-generation. Gend Dev 17(1):51–66
CARE (2010) Adaptation, gender and women's empowerment. CARE International Climate Change Brief. CARE International, Washington, DC. http://www.careclimatechange.org/files/adaptation/CARE_Gender_Brief_Oct2010.pdf. Accessed 30 Jan 2012
Carr E (2008) Between structure and agency: livelihoods and adaptation in Ghana's central region. Glob Environ Chang 18(4):689–699
Conway T, Moser C, Norton A, Farrington J (2002) Rights and livelihoods approaches: exploring policy dimensions. Nat Resour Perspect 78:1–6
Cornwall A, Molyneux M (2006) The politics of rights: dilemmas for feminist praxis: an introduction. Third World Q 27(7):1175–1191
Cornwall A, Nyamu-Musembi C (2004) Putting the 'rights-based approach' to development into perspective. Third World Q 25(8):1415–1437
Cornwall A, Harrison E, Whitehead A (2007) Gender myths and feminist fables: the struggle for interpretive power in gender and development. Dev Chang 38(1):1–20
Cuomo C (2011) Climate change, vulnerability, and responsibility. Hypatia 26(4):690–714
Dankelman I (2010) Introduction: exploring gender, environment, and climate change. In: Dankelman I (ed) Gender and climate change: an introduction. Earthscan, London, pp 1–20
Davies M, Oswald K, Mitchell T (2009) Climate change adaptation, disaster risk reduction and social protection. In: Promoting pro-poor growth: social protection, Organization for Economic Co-operation and Development, pp 201–217. http://www.oecd.org/dataoecd/25/55/43280946.pdf. Accessed 30 Jan 2012
Devereux S (2006) Cash transfers and social protection. Institute of Development Studies, Brighton. http://sarpn.org/documents/d0002199/Cash_Transfer_Devereux_Oct2006.pdf. Accessed 30 Jan 2012
Devereux S, Sabates-Wheeler R (2004) Transformative social protection. IDS working papers 232, Institute of Development Studies, Brighton. www.ids.ac.uk/files/Wp232.pdf. Accessed 30 Jan 2012
Drinkwater M (2009) "We are also human": identity and power in gender relations. In: Hickey S, Mitlin D (eds) Rights-based approaches to development: exploring the potential and pitfalls. Kumarian Press, Sterling, pp 145–164
Ellis F, Devereux S, White P (2009) Social protection in Africa. Edward Elgar, Cheltenham
Folke C, Carpenter S, Elmqvist T, Gunderson L, Holling C (2002) Resilience and sustainable development: building adaptive capacity in a world of transformations. AMBIO J Hum Environ 31(5):437–440
Glazebrook T (2011) Women and climate change: a case-study from northeast Ghana. Hypatia 26(4):762–782

Goonesekere S (1998) A rights-based approach to realizing gender equality. UN Division of the Advancement of Women, New York. http://www.un.org/womenwatch/daw/news/savitri.htm. Accessed 30 Jan 2012

Heltberg R, Siegerl PB, Jorgenson SL (2009) Addressing human vulnerability to climate change: toward a 'no-regrets' approach. Glob Environ Chang 19:89–99

Hernández Castillo RA (2002) National law and indigenous customary law: the struggle for justice of indigenous women in Chiapas, Mexico. In: Molyneux M, Razavi S (eds) Gender justice, development, and rights. Oxford University Press, Oxford, pp 384–412

Hickey S, Mitlin D (2009) Introduction. In: Hickey S, Mitlin D (eds) Rights-based approaches to development: exploring the potential and pitfalls. Kumarian Press, Sterling, pp 3–20

Lambrou Y, Piana G (2006) Gender: the missing component of the response to climate change. Food and Agricultural Organization of the United Nations, Rome. http://www.eldis.org/vfile/upload/1/document/0708/DOC21057.pdf. Accessed 30 Jan 2012

Li T (2007) The will to improve: governmentality, development, and the practice of politics. Duke University Press, Durham

MacGregor S (2009) A stranger silence still: the need for feminist social research on climate change. Sociol Rev 57:124–140

MacGregor S (2010) Gender and climate change: from impacts to discourses. J Indian Ocean Reg 6(2):223–238

Manzo K (2010) Imaging vulnerability: the iconography of climate change. Area 42(1):96–107

Masika R (2002) Editorial. Gend Dev 10(2):2–9

McIlwaine C, Datta K (2003) From feminising to engendering development. Gen Place Cult 10(4):369–382

Mohanty CT, Russo A, Torres L (1991) Third world women and the politics of feminism. Indiana University Press, Indiana

Molyneux M, Lazar S (2003) Doing the rights thing: rights-based development and Latin American NGOs. ITDG Publishing, London

Molyneux M, Razavi S (2002) Gender justice, development, and rights. Oxford University Press, Oxford

Moraga C, Anzaldua G (1984) This bridge called my back: writings by radical women of color. Women of Color Press, New York

Moser C, Norton A (2001) To claim our rights: livelihood security, human rights and sustainable development. Overseas Development Institute, London

Nelson V (2011) Gender, generations, social protection and climate change: a thematic review. Overseas Development Institute, London

Nelson V, Stathers T (2009) Resilience, power, culture, and climate: a case study from semi-arid Tanzania, and new research directions. Gend Dev 17(1):81–94

Nightingale A (2009) Warming up the climate change debate: a challenge to policy based on adaptation. J For Livelihood 8(1):84–89

O'Brien K, Leichenko R, Kelkar U, Venema H, Aandahl C, Tompkins H, Javed A, Bhadwal S, Barg S, Nygaard L, West J (2004) Mapping vulnerability to multiple stressors: climate change and globalization in India. Glob Environ Chang 14:303–313

O'Brien K, Sygna L, Leichenko R, Adger WN, Barnett J, Mitchell T, Schipper L, Tanner T, Vogel C, Mortreux C (2008) Disaster risk reduction, climate change adaptation and human security. Global Environmental Change and Human Security Project, Oslo. http://connectgreen.com.br/site/images/online/arquivo_GECHS_Report_3-08.pdf. Accessed 30 Jan 2012

Ong A (1988) Colonialism and modernity: feminist representations of women in non-Western societies. Inscriptions 3/4:79–104

Onta N, Resurreccion BP (2011) The role of gender and caste in climate adaptation strategies in Nepal. Mt Res Dev 31(4):351–356

Oxfam (2009) People-centred resilience: working with vulnerable farmers towards climate change adaptation and food security. Oxfam briefing paper 135, Oxfam International, Oxford. http://www.oxfam.org/sites/www.oxfam.org/files/people-centred-resilience-en-0911.pdf. Accessed 30 Jan 2012

Oxfam & United Nations Vietnam (2009) Responding to climate change in Viet Nam: opportunities for improving gender equality. A policy discussion paper, Oxfam International and UN Vietnam, Ha Noi, Vietnam. http://www.oxfam.org.uk/resources/policy/climate_change/downloads/climate_change_gender_equality_vietnam.pdf. Accessed 30 Jan 2012

Petrie B (2010) Gender and climate change: regional summary. Heinrich Boll Foundation Southern Africa, Cape Town

Pettengell C (2010) Climate change adaptation: enabling people living in poverty to adapt. Oxfam International Research Report. http://www.mountainforum.org/sites/default/files/pub/6365.pdf. Accessed 30 Jan 2012

Practical Action (2010) Integrating approaches: sustainable livelihoods, disaster risk reduction, and climate change adaptation. Practice briefing, Practical Action, Warwickshire. http://www.preventionweb.net/english/professional/publications/v.php?id=13978. Accessed 30 Jan 2012

Radcliffe S (2002) Indigenous women, rights and the nation-state in the Andes. In: Craske N, Molyneux M (eds) Gender and the politics of rights and democracy in Latin America. Palgrave, Basingstoke, pp 149–161

Ray-Bennet NS (2010) The role of microcredit in reducing women's vulnerabilities to multiple disasters. Disasters 34(1):240–260

Resurreccion BP (2011) The gender and climate change debate: more of the same or new pathways of thinking and doing? Asia security initiative policy series, 10. RSIS Centre for Non-Traditional Security Studies, Singapore

Robinson M (2005) What rights can add to good development practice? In: Alston P, Robinson M (eds) Human rights and development: towards mutual reinforcement. Oxford University Press, New York, pp 25–44

Rossi A, Lambrou Y (2008) Gender and equity issues in liquid biofuels production: minimizing the risks to maximize the opportunities. Food and Agriculture Organization, Rome. ftp://ftp.fao.org/docrep/fao/010/ai503e/ai503e00.pdf. Accessed 30 Jan 2012

Schipper L (2009) Meeting at the crossroads? Exploring the linkages between climate change adaptation and disaster risk reduction. Clim Dev 1:16–30

Segnestam L (2009) Division of capitals: what role does it play for gender-differentiated vulnerability to drought in Nicaragua? Community Dev 40:154–176

Smit B, Pilifosova O (2003) From adaptation to adaptive capacity and vulnerability reduction. In: Smith BT, Klein RJT, Huq S (eds) Climate change, adaptive capacity and development. Imperial College Press, London, pp 9–28

Smit B, Wandel J (2006) Adaptation, adaptive capacity and vulnerability. Glob Environ Chang 16:282–292

Stirbu M (2010) Climate change and social protection in Cambodia: linkages and synergies. United Nations Children's Fund, New York. http://www.eldis.org/assets/Docs/56243.html. Accessed 30 Jan 2012

Symington A (2002) A rights-based approach to development. Association Women's Rights Dev Women's Rights Econ Chang 1:1–7

Tandon N (2007) Biopolitics, climate change and water security: impact, vulnerability and adaptation issues for women. Agenda 27:4–20

Terry G (2009) No climate justice without gender justice: an overview of the issues. Gend Dev 17(1):5–18

Tschakert P, Dietrich K (2010) Anticipatory learning for climate change adaptation and resilience. Ecol Soc 15(2):11–33

Tschakert P, Sagoe R, Darko G, Codjoe SN (2010) Floods in the Sahel: an analysis of anomalies, memory, and anticipatory learning. Clim Chang 103(3–4):471–502

Walby S (2005) Gender mainstreaming: productive tensions in theory and practice. Soc Polit Int Stud Gend State Soc 12(3):321–343

Woodford-Berger P (2004) Gender mainstreaming: what is it (about) and should we continue doing it? IDS Bull 35(4):65–72

Chapter 8
From "Free" Trade to Farm Women: Gender and the Neoliberal Environment

Amber J. Fletcher

Abstract This chapter challenges the economism of contemporary macroeconomic policy through an environmental and gendered lens. It uses a feminist political economy framework to analyze newly released statistical data on international trade flows from the Food and Agriculture Organization (FAO), as well as statistics on Canadian agricultural production. The analysis moves through three levels of inquiry – macro, meso, and micro – to examine the interaction of neoliberal policy, gender, and environment. At the macro level, it questions the neoliberal discourse of "efficiency" by using trade data to illustrate the environmental inefficiencies of the current trade regime, especially the phenomenon of "trade for trade's sake". It links the neoliberal policy regime to its environmental consequences at the meso level of Canadian agricultural production. Finally, the gendered effects of neoliberalism are illustrated by examining the situation of Canadian farm women. Although macro-level policies are often portrayed as "gender neutral", their unique effects on farm women's lives are elucidated using the concept of social reproduction.

Keywords Gender • Free trade • Farm women • Neo-liberal environment • Canada

This chapter is based on a paper presented at the "Gender and Climate Change: Women, Research, and Action" conference, 16 September 2011, at Monash University in Prato, Italy. The author gratefully acknowledges the support of the Social Sciences and Humanities Research Council of Canada (SSHRC) and the Regina Public Interest Research Group (RPIRG).

A.J. Fletcher (✉)
Canadian Plains Studies, University of Regina, #110.1 – 2 Research Drive, Regina, SK S4S 0A2, Canada
e-mail: amber.fletcher@uregina.ca

M. Alston and K. Whittenbury (eds.), *Research, Action and Policy: Addressing the Gendered Impacts of Climate Change*, DOI 10.1007/978-94-007-5518-5_8,

In the current context of global trade and global climate change, there is tension between dominant economic and ecological value systems. While some theorists have challenged the self-regulating market for its derogatory environmental and social effects, some theorists have argued that the self-regulating market is inherently good for the environment. This chapter challenges the narrow economism of this latter vision by illustrating its environmental and human consequences at three levels of inquiry: macro, meso, and micro.

This multi-level approach exposes the concrete environmental effects of macroeconomic policies, while also elucidating the uniquely gendered impacts and adaptations that may only be visible at the micro level of production and social reproduction. The situation of Canadian farm women provides an example of gendered micro-level adaptation at the confluence of economic and environmental challenges. It allows us to understand the gendered and environmental consequences of macroeconomic policies that are often seen, superficially, as gender neutral or even environmentally beneficial.

At the macro level, this chapter discusses the marginalization of environmental concerns during the creation of the North American Free Trade Agreement (NAFTA). The environmental consequences of such neoliberal trade regimes have become evident in the unnecessary movement of goods across borders. At the meso level of domestic economic organization, trade agreements have opened a door to the increasing corporatization of Canadian agriculture, which has its own discrete environmental effects in the form of air and soil degradation. In the Canadian context, the gendered effects of neoliberal macroeconomics become truly visible by lowering the scale of analysis to the micro level. Through the lens of gendered relations of production and social reproduction, it becomes clear that women on family farms are expanding their labour capacity as an adaptive response to broader economic and environmental stressors.

The purpose of this chapter is not to offer a comprehensive analysis of environmental damage caused by "free trade" regimes. Rather, it poses a challenge to pro-trade perspectives that reify the environmental and human benefits of trade. It challenges the neoliberal discourse of trade "efficiency" by highlighting some particular environmental inefficiencies, as well as social deficiencies, associated with the current neoliberal system of global trade.

The Macro Level

The Neoliberal Economic "Environment"

Over the past four decades, the world economy has witnessed a dramatic shift toward international economic integration. From its gradual beginnings in the 1947 General Agreement on Trade and Tariffs (GATT), economic integration has accelerated rapidly since the 1970s. The elimination of the Bretton Woods system and the resulting commoditisation of currency contributed to highly globalized systems of production and finance. Many states implemented monetarist economic policies, and international trade was seen as the key to domestic growth. The 1980s and 1990s were marked by a spate of free trade agreements, such as the 1995 North American Free Trade Agreement (NAFTA), and also saw the 1995 creation of the World Trade Organization.

International trade is not just a contemporary phenomenon in Canada (McBride 2001). The country's colonization-era policies were intended, in part, to establish the prairies as a site of export-oriented agricultural production (Fowke 1957). However, today's trade relations are shaped by the unique contemporary context of global financialization and neoliberal governance. Neoliberal ideology purports that "human well-being can best be advanced . . . within an institutional framework characterized by strong private property rights, free markets, and free trade" (Harvey 2005, p. 2).

Political theorists have challenged the accuracy of the notion "free" trade. It is argued that, rather than diminishing entirely, the role of the state has in fact been *reoriented* to encourage trade and foreign investment (Harvey 2005; Petras and Veltmeyer 2001), while simultaneously withdrawing from some forms of regulation and social provisioning (Bakker 2003; Luxton 2007). Indeed, Panitch and Gindin (2009) have argued that neoliberalism must not be viewed only as a governing ideology, but also as a form of social rule in which states actively facilitate the creation of new capitalist markets. They argue that "it is more important than ever to distinguish between understanding neoliberalism as an ideologically driven strategy to free markets from states, as opposed to a materially driven form of social rule that has involved the liberalization of markets through state intervention and management". One example of such intervention is what Elson and Cagatay (2000) have called the "deflationary" policy bias, wherein "liberalized financial markets have induced governments to adopt policies aimed at maintaining their 'credibility' in financial markets", through fiscal restraint measures, for example (p. 1354).

Stephen Gill (2005) points out the social consequences of this paradigm, stating that "fiscal crisis and cutbacks in government expenditures to service debts affects communities unequally. The poorest tend to be the hardest hit – often women, children and the aged – when provisions for social reproduction are removed or reduced" (p. 128). Further, it is not just the removal of social programs that can have negative social consequences. It is necessary to illustrate the social and environmental consequences of policies that do not always appear, at the outset, to have direct implications for society or environment. Free trade agreements are one example.

Neoliberalism, "Free" Trade, and Environment

From a neoliberal perspective, free trade is often treated as environmentally unproblematic, possibly because trade is accepted as entirely necessary and inherently efficient. The major free trade agreements of the mid-1990s came on the heels of several major international environmental commissions and agreements, such as the 1987 World Commission on Environment and Development, which produced the Brundtland Report, and the 1992 United Nations Conference on Environment and Development. Despite the increased awareness of environmental concerns at that time, the agreements addressed these concerns in only a peripheral way. In some cases, free trade was simply assumed to be *conducive to* environmental preservation. A 1999 World Trade Organization (WTO) report, for example, stated that environmental policies should foster an increasingly integrated global economy, "*so as to ensure* that we live within ecological limits" (p. 7, emphasis added) (World Trade Organization 1999).

Despite Kirton's (2002) assertion that NAFTA contains an "embedded ecologism" (Conclusion, para. 1), the construction of the NAFTA illustrates the submersion of environmental concerns beneath the economic juggernaut. In 1993, the increased global environmental awareness had created a dilemma for the economically focused NAFTA document. The NAFTA, as it stood upon completion that year, "did not address key public concerns in the areas of labour or the environment" (Carpentier 2006, p. 259). This meant that US President Bill Clinton faced challenges in obtaining congressional approval of the agreement. It proved necessary to create "side agreements" on environment to accompany the NAFTA, in order to address these concerns (Carpentier 2006, p. 259). These side agreements were accepted as an alternative to "clause-by-clause congressional review" of the NAFTA (Carpentier 2006, p. 259), which would have required examining the agreement more closely for its potential environmental impacts. Thus, the side agreements constitute what may be called a "quick fix" to the lack of environmental consideration in the main text.

The side agreements on environment became the North American Agreement on Environmental Cooperation (NAAEC), which was administered through the newly created Commission for Environmental Cooperation (CEC). According to Kirton (2002), the CEC remains underfunded and, although conducting very in-depth investigations, is mandated only to "*monitor* the environmental effects" of the NAFTA after they have already occurred (Carpentier 2006, p. 259, emphasis added). When considering potentially irreversible ecological consequences, this type of *ex-post* monitoring is problematic. Perhaps more problematic is the fact that, according to a 2006 paper by a CEC official, the CEC "was not mandated" to collect empirical environmental data prior to the onset of NAFTA (Carpentier 2006, p. 259); thus, "data availability has frustrated [its] ability to conduct comprehensive modelling analysis of the environmental effects of NAFTA" (Carpentier 2006, p. 264). This clearly illustrates the marginal position of environmental concerns in NAFTA, particularly vis-à-vis the emphatic focus on economic issues.

The privileging of economics over environment is also manifest in the actual practice of trade. Although international trade has occurred for centuries, contemporary trade relations are marked by an intensive approach that highlights the internationalization of production and the globalization of finance. In the case of production, Shields and Evans (1998) have discussed the "global factory", wherein a single product can be comprised of pieces sourced from many parts of the world. This is a phenomenon rooted in the Ricardian notion of competitive advantage.

On the consumption side, the combined neoliberal forces of financialization and neoliberal trade have brought about a troubling trend: food trade that is largely unnecessary in practical terms, but which is conducted primarily for the maximization of profit, by taking advantage of financial market factors. Statistics from the Food and Agriculture Organization clearly illustrate this trend.

Trade for Trade's Sake?

In 2009, Canada exported 304,048 tonne of boneless beef to other countries (FAO 2009). In the very same year, Canada also imported 132,702 tonne of the same commodity (FAO 2009). Thus, in 2009, boneless beef was Canada's eighth most valuable *export* and also its fifth most valuable *import* (FAO 2009). That year, Canada was the world's tenth ranking exporter of boneless beef, and also the world's 12th ranking importer of boneless beef (FAO 2009). This situation also exists for pork. In 2009, Canada exported 658,765 tonne of pork across its borders; the same year, it imported 110,774 tonne of the same commodity (FAO 2009).

Clearly, this trade did not occur because Canada is unable to produce beef and pork. Because the goods were exported and imported within the same category of processing, it cannot be deduced that these products were exported for the purpose of substantial processing. Because the Canadian and American beef and pork industries are concentrated within specific regions of each country, and because of the vast geographical expanse of the two nations, some Canada-USA trade may actually reduce physical transport distance. That is, some locations in the United States may be closer destinations for Canadian shipments than some Canadian destinations. However, other international examples show that not all international food trade is efficient.

For example, in 2009, the year that Canada exported 304,048 tonne of boneless beef worldwide, it imported 29,921 tonne of boneless beef from New Zealand, a country that is certainly not geographically convenient to Canada (FAO 2009). That same year, Canada also exported 1,936 tonne of pork to Chile, while simultaneously importing 2,338 tonne of the same product from the same country (FAO 2009). As another example, in 2009 Australia exported 162,412 tonne of whole-milk cheese to 63 countries worldwide, and simultaneously imported 59,045 tonne of the same product that year from 28 different countries (FAO 2009). Several of these countries, including New Zealand and the United States, both received cheese from Australia and also sent cheese to it.

As Norberg-Hodge and colleagues have argued, this "excessive transport" in food products "benefits only a few large-scale agribusinesses and speculators, which take advantage of... exchange rate swings and price differentials to shift foods from country to country in search of the highest profits" (2002, 19). Due to increasing vertical integration in the agricultural sector, many of the same agribusinesses involved in export are also involved in both production and processing (Desmarais 2007; Novek 2003). These corporations benefit from neoliberal trade policies because trade liberalization allows an influx of imported products into domestic markets, driving down prices. Although some corporations are engaged directly in production, others choose instead to contract with farmers to provide livestock for processing. In the case of the hog sector, Qualman (2001) has pointed out that large agribusinesses benefit from lowered product prices because this "allow[s] them to take their profits in their packing plants and to secure additional hog supplies from independent producers at low cost" (p. 26).

Contemporary trade relations have also facilitated the movement of agricultural products across borders for the purpose of processing. This includes the long-distance transport of live animals for slaughter, processing, and packing. In 2006, Canada exported 8,776,985 live hogs from 8 of its provinces (Statistics Canada, Table 1, Table 15, n.d. a, b). The majority of these hogs were shipped to the United States for slaughter and processing, with many returned to Canada in the form of imported pork – 78,347 tonne in 2006 (FAO 2006). Other top destinations for Canadian hogs include Mexico, Russia, and Chile, while some hogs are shipped as far as the Philippines and South Korea (Statistics Canada, Table 23, n.d. c). Although Canada sent 1,804 live hogs to Chile in 2006 (Statistics Canada, Table 23, n.d. c), it also imported hogs in their processed form – pork – from this country. A similar relationship, although on a smaller scale, exists with Denmark.

This long-distance trade movement poses concerns about humane animal treatment, as well as environmental consequences. Long-distance transport for the purpose of processing can result in diminishing domestic processing capacity, guided by a philosophy of comparative advantage. Indeed, in a paper presented at a hog industry conference in 2006, Orr and Shen point out that "this large export trade in pork and live animals reflects the response of the North American pig industry to it's [*sic*] competitive advantages (Young 2005). Canada has better herd reproductive efficiency than the USA, but the U.S. is very competitive in finishing pigs for market, along with a generally stronger American dollar so as to purchase pigs at a very competitive price" (n.p.). Thus, Ricardian notions of comparative advantage combine with a neoliberal trade ideology to ultimately perpetuate the cycle of "long distance" processing.

In the case of Canada-USA trade in live animals, trucking is a key mode of transport. The NAFTA has dramatically increased the use of trucking for trade. In 1989, the USA imported one million live Canadian hogs; by 2004, the number had increased to eight million (Haley 2004). Similarly, the Sierra Club and Holbrook-White (2000) point out that, in general, "with the convenience of door-to-door delivery and the increasing use of just-in-time inventory controls, trucks have emerged as the leading mode of transport for NAFTA freight as measured by the value of trade" (p. 9).

All this trade movement may prove economically "efficient" by resulting in higher profits for agri-food processor-exporters. However, the environmental consequences of unnecessary trade are rarely factored into trade economics, and are seen instead as externalities. According to the CEC (2011), trucking produces 104.5 tonne of carbon dioxide per million-tonne kilometre shipped, and cargo shipping produces 28.2 tonne. Although both release fewer carbon emissions than air transport, the key point remains that such trade is ultimately unnecessary in practical terms and, as such, some of these emissions could be eliminated entirely if not for the corporate profit motive.

The Meso Level

The Agricultural "Environment": Corporatization and the Family Farm

At the meso level, the environmental consequences of neoliberal economics are visible not only in the way agricultural products are transported, but also in how they are produced. The Canadian agricultural landscape is shifting toward an increasingly large-scale, industrialized mode of production. Between 2001 and 2006, the number of farms with less than $250,000 in receipts, which are typically family farms,[1] declined by 10.5% (Statistics Canada 2006a). In contrast, the number of farms with at least $1 million in receipts, which are typically large-scale corporate farms, jumped 32.5% (Statistics Canada 2006a). Although these large corporate farms account for only 2.6% of farms in the production sector, they receive 40% of total farm income (Statistics Canada 2006a).

The shift has been facilitated and enabled by neoliberal governmental policy at the domestic level. Throughout the 1980s and 1990s, the federal government began to dismantle support programs for domestic producers, often in response to the imperatives of free trade agreements. One example is the elimination of supply management mechanisms, which were created to provide more stable prices for producers. In 1996, for example, the provincial government of Manitoba eliminated the provincial supply-management mechanism for hogs. The price of hogs, which had averaged $1,240 per tonne between 1991 and 1997, dropped to $697 per tonne by 1998. Pork producers had still seen only a gradual rise to $1,027 by 2008 (FAO 2008). This, combined with a tripling of pork imports, caused the number of hog farms to drop by 54% by 2001, as many family producers found themselves

[1] Stirling (2001) has defined the "family farm" as "a type of farming in which families own all or most of their productive capital – land, buildings, machinery, and livestock – and do all or most of the productive work themselves. Typically they live in the communities where they farm, their farms are not large, and there are many of them" (p. 248). In a similar vein, Furtan has defined "commercial farms" in opposition to the family farm, because in a commercial farm "the operator is engaged in less of the ownership, management, risk and/or reward of the farming operation even though farm operators and their families continue to provide the labour to the agricultural production operation" (Furtan 2006, p. 10).

unable to survive in the economic environment (Tait 2003). Today, pig production in Manitoba is concentrated in large-scale industrial operations (Novek 2003), which provide low-cost inputs to processing corporations (Qualman 2001).

Dominant economic discourse extols this type of large-scale industrial agriculture as a way to "feed the world". This mantra is short sighted, failing to consider that the average life span of an industrial hog facility is only 15–20 years (Ervin et al. 2003). It also fails to consider the long-term environmental consequences of industrial-scale agricultural production. Whereas smaller-scale mixed farms can use animal manure productively as a fertilizer, intensive livestock operations (ILOs) often house tens of thousands of animals in close conditions. In addition to the increased risk of disease proliferation and inhumane treatment of animals, these farms produce more waste than they can sustainably process using their small land base (Thu 2003). In the hog industry, this results in large cesspools of liquefied manure, which is then spread upon or injected into the land in quantities that are destructive to soil quality, and contaminates groundwater with excess nitrogen and phosphorus (Thu 2003; Tait 2003).

In contrast to the industrialized system, past research has shown that family farmers possess an environmental awareness, and a desire to preserve the land over the long-term (Lind 1995). In the American context, Horrigan et al. (2002) illustrated the environmental advantages of the "family farm" model over the industrial livestock operation, reporting that "the average US farm uses 3 kcal of fossil energy in producing 1 kcal of food energy (in feedlot beef production, this ratio is 35:1)" (p. 446). They also point out that grain used in supply feedlots is grown in intensive monocultures, which can also have negative environmental effects (p. 445). However, due to the pressure to survive in an increasingly competitive environment, many family farmers now find themselves in a position of necessary productivism, which forces them to use environmentally damaging strategies to increase production and stay competitive.

The Micro Level

The Gendered "Environment": Farm Women and Sustainability

Milne (2005) has argued that "in wealthier societies, like Canada, there is likely to be less gender differentiation from the effects of climate change" than in poorer countries (p. 51). As such, it can be difficult to isolate the specifically gendered effects of the phenomenon. Similarly, due to the often-perceived degrees of separation between macroeconomic policies and people's everyday lives, the gendered effects of these policies are not often readily comprehensible. However, by applying a gendered analysis to the everyday conditions of the neoliberal paradigm, it is possible to see that Canadian farm women are specifically and uniquely affected by the interaction of neoliberalism and climate change. This becomes apparent through the lens of social reproduction.

Bakker and Silvey (2008) have described social reproduction as "both biological reproduction of the species (including its ecological framework) and on-going reproduction of the commodity labour power" (p. 2). As such, social reproduction is not peripheral to the mode of production; rather, it is "at its centre" (Bakker and Silvey 2008, p. 2). Early feminist political economy focussed primarily on "value surplus creation" through women's unwaged domestic labour; however, Bakker and Silvey argue further that relations of social reproduction shape, and are shaped by, the broader political-economic order (Bakker and Silvey 2008).

This point is illustrated by examining gender relations on Canadian family farms. As previously mentioned, Canadian family farmers are experiencing the effects of intertwining corporatization and government re-regulation in their sector. Although farm families are profoundly affected by these changes, Phillips (1998) has challenged the theoretical tendency to reify macro-level forces as entirely structuring. As she and Martz (2006) have pointed out, farm families are not entirely passive to macro-level forces; rather, they respond to changes through particular adaptive strategies at the micro level. For example, Martz's (2006) research illustrated that farm families have diversified their activities as an adaptive strategy in the face of financial stress. Martz views these individuals as active agents who, as a family unit, negotiate and respond to the changes they experience. Both authors avoid reification and homogenization of the "farm family", noting that these negotiations take specifically gendered forms.

On the Canadian prairies, two main adaptive responses to economic stress have been to increase the size of the farm and/or supplement farm labour with off-farm employment. Farm women are increasing their labour in both these areas, while strongly rooted gender ideologies also ensure their disproportionate responsibility for domestic and caregiving labour. As a result, farm women's labour is increasingly stretched in a number of different directions, both on and off the farm, producing what Wiebe (1995) has identified as the "triple day".

Canadian farmers, especially farm women, are now working in off-farm or non-farm employment at an unprecedented level (Martz 2006). Currently, 47.7% of farm men and 50.4% of farm women work off the farm (Statistics Canada 2006b). One study reported that for farm women, off-farm work has increased by over 50% between 1982 and 2002, and their rates of full-time work have now surpassed that of farm men, at 34% compared to 29% for men (Martz 2006). Studies on farm women (Kubik 2004; Martz 2006) have shown that the reasons for their off-farm employment were overwhelmingly financial. Indeed, off-farm labour now contributes to approximately half of total farm income (Statistics Canada 2006b); thus, it acts as a crucial support for farming families facing economic challenges.

In addition to increasing off-farm work, Canadian farm women are also dramatically increasing their level of farm labour in all areas (Martz 2006). A longitudinal study found that, between 1982 and 2002, farm women's participation in fieldwork increased by 12%, and their farm management activities increased by 22% (Martz 2006). This increase is, at least in part, attributable to the growing size and complexity of Canadian family farms, as farmers struggle to survive in a difficult economy and cope with increased farm management work. In some cases, farm women assume the role of hired workers as farms become unable to pay these hired workers (Alston 1998). Nonetheless, farm life is still marked by a gendered division of labour. On the farm, men are still conceptualized as the central "farmers" whose work is most likely to be seen as "farm" work (Kubik 2004; Martz 2006), whereas women often play the role of assistants or "go-fers" whose own work is side-lined and often interrupted by "farm" tasks (Faye 2006).

This gendered division of labour continues to structure relations of social reproduction. Studies have shown that Canadian farm women perform an overwhelming majority of caregiving and domestic work in rural areas (Jaffe and Blakley 1999; Kubik 2004). In the rural context, this work is complicated by a lack of institutional supports for caregiving (Jaffe and Blakley 1999), as well as the complexity of domestic tasks on the farm, where "cooking" can mean feeding numerous hired workers and "laundry" often involves handling clothes covered in hazardous chemicals (Kubik 2004). However, such work is rarely viewed as "farm" work in the same way as seeding or harvesting.

Women's unwaged work also extends to the broader community. Jaffe (2003) pointed out that many farm women also play an important role in building and maintaining rural communities through volunteer work. In fact, 82% of Martz's (2006) respondents were active as volunteers in their community, despite the growing demand for their time in other areas. Her 2006 research showed that farm women performed more volunteer work than farm men. Further, recent changes in rural infrastructure, such as the closure of hospitals and long-term care homes, have resulted in an increased reliance on the unwaged caregiving labour of women (Jaffe and Blakely 1999; Luxton 2007), a situation that is exacerbated by the aging demographic trend in rural Canada. As such, neoliberal policies are often premised on the assumption that rural women's labour will simply stretch to compensate the offloading of government responsibility for care services.

Feminist scholars have discussed the continued invisibility of farm women's unwaged contributions, both on and off the farm, and the consequences of this invisibility (Alston 1998; Kubik 2004; Kubik and Fletcher forthcoming). This is enabled by a "divisive spatial imagination" (Whatmore 1991, 145), which rigidly divides family from economy, household from farm. This division has marginalized the household as somehow peripheral to the farm, and farm women's contributions as secondary to the "productive" farm activities done primarily by men. Alston (1998) describes this as "the tradition of devaluing the labour contribution of women and of discounting their economic contributions to the farm production process" (p. 24). Because gendered power relations at the micro-level often go unseen, so does the interaction of these relations with macro-level economic policies, which are also discursively constituted as gender neutral. Decreased support levels for family farmers in Canada mean that farmers will continue to respond through the extension and diversification of their labour capacity – and through the continued "stretching" of farm women's invisible, unwaged labour in a number of directions.

Economic challenges in the agriculture sector are soon to be compounded by increased environmental challenges. Climate scientists observe a general warming trend that will result in increased incidents of climate extremes in the Canadian prairie region, particularly floods and droughts (Sauchyn and Kulshreshtha 2008). In 2009, areas of Saskatchewan experienced drought, which was immediately followed by severe flooding in parts of the prairie provinces during 2010 and 2011. Total government expenditures for one such flood event, which hit southern Alberta and Saskatchewan in June of 2010, amounted to $956,350,000 and resulted in the evacuation of 2,065 people (Public Safety Canada, n.d.). Research in the Australian context (Alston 2006) has indicated that entrenched gender ideologies are both reinforced and reshaped in the context of climate extremes. During drought events, Australian farm women assumed "breadwinner" roles both on- and off-farm, yet also maintained a strong caregiving role by ensuring the emotional stability of family members under stress (Alston 2006), both of which indicate the extension of farm women's existing roles. Further research is necessary in the Canadian context to explore the interaction between recent climate events and farm women's work as an adaptive strategy.

This chapter has used a multi-level framework to elucidate the connections between macroeconomic trade policies, environmental damage, and their gendered human consequences. Neoliberal policy frameworks emphasize economic wellbeing over environmental concerns, resulting in tonnes of transport-related emissions that ultimately provide little human benefit. The very same frameworks have enabled a re-orientation of agricultural policy toward an increasingly industrialized and corporatized model, which carries its own environmental consequences. As a site where the consequences of neoliberalism are strongly felt, micro-level social relations – such as family farm production and social reproduction – have become a key site of adaptation to the consequences of neoliberalism.

On the family farm, farmers exploit their own land and labour to keep the farm alive in the face of increasing economic and environmental challenges, with farm women constituting an under-recognized, often invisible labour source. It is necessary to illustrate the centrality of gender and farm women's labour (waged and unwaged) to current systems of production, in order to draw broad linkages between macroeconomic policy and its gendered effects. Such an analysis elucidates the human and environmental "externalities" of the neoliberal economic system. It urges us to centralize and prioritize not only environmental sustainability, but also sustainability in a human sense, over narrow economism.

References

Alston M (1998) Farm women and their work: why is it not recognized? J Sociol 34(1):23–34

Alston M (2006) The gendered impact of drought. In: Bock BB, Shortall S (eds) Rural gender relations: issues and case studies. CABI, Cambridge, pp 165–180

Bakker I (2003) Neo-liberal governance and the reprivatization of social reproduction: social provisioning and shifting gender orders. In: Bakker I, Gill S (eds) Power, production and social reproduction: human in/security in the global political economy. Palgrave McMillan, Basingstoke/New York, pp 66–83

Bakker I, Silvey R (2008) Introduction: social reproduction and global transformations – from the everyday to the global. In: Bakker I, Silvey R (eds) Beyond states and markets: the challenges of social reproduction. Routledge, London/New York, pp 1–15
Carpentier CL (2006) NAFTA commission for environmental cooperation: ongoing assessment of trade liberalization in North America. Impact Assess Proj Apprais 24(4):259–272
Commission for Environmental Cooperation (CEC) (2011) Destination sustainability: reducing greenhouse gas emissions from freight transportation in north America. CEC, Montreal
Desmarais A (2007) La via campesina: globalization and the power of peasants. Fernwood, Halifax
Elson D, Cagatay N (2000) The social content of macroeconomic policies. World Dev 28(7):1347–1365
Ervin AM, Holtslander C, Qualman D, Sawa R (2003) Introduction. In: Ervin AM, Holtslander C, Qualman D, Sawa R (eds) Beyond factory farming: corporate hog barns and the threat to public health, the environment, and rural communities. Canadian Centre for Policy Alternatives, Saskatoon, pp 1–8
Faye L (2006) Redefining 'farmer': agrarian feminist theory and the work of Saskatchewan farm women. (Unpublished Master thesis), Memorial University of Newfoundland, St. John's
Food and Agriculture Organization (FAO) (2006, 2008, 2009, 2011) FAOStat. Retrieved August 3, 2011, from http://faostat.fao.org/site/342/default.aspx and http://faostat.fao.org/site/537/DesktopDefault.aspx?PageID=537
Fowke V (1957) The national policy and the wheat economy. University of Toronto Press, Toronto
Furtan H (2006) The need for direction: the Canadian grains sector at a crossroads. Knowledge Impact in Society Publications, Saskatoon
Gill S (2005) Conceptualizing the global restructuring of public services: privatization, TRIPS & GATS. Jurid Z im Rechtsstaat 4(3):128–134
Haley MM (2004) Market integration in the North American hog industries. United States Department of Agriculture, LDP-M-125-01
Harvey D (2005) A brief history of neoliberalism. Oxford University Press, Oxford
Horrigan L, Lawrence RS, Walker P (2002) How sustainable agriculture can address the environmental and human health harms of industrial agriculture. Environ Health Perspect 110(5):445–456
Jaffe J (2003) Family labour processes, land, and the farm crisis in Saskatchewan. In: Diaz HP, Jaffe J, Stirling R (eds) Farm communities at the crossroads. Canadian Plains Research Center, Regina, pp 137–148
Jaffe J, Blakley B (1999) Coping as a rural caregiver: the impact of health care reforms on rural women informal caregivers. Prairie Women's Health Centre of Excellence, Regina
Kirton J (2002) NAFTA's trade-environment regime and its commission for environmental cooperation: contributions and challenges ten years on. Can J Reg Sci 25(2) n.p
Kubik W (2004) The changing roles of farm women and the consequences for their health, well being, and quality of life. (Unpublished doctoral dissertation), University of Regina, Regina
Kubik W, Fletcher A (Forthcoming) The multiple dimensions of health: weaving together food sustainability and farm women's health. In: Leipert B, Leach B, Thurston W (eds.) Rural women's health. University of Toronto Press, Toronto, pp 185–209
Lind C (1995) Something's wrong somewhere: globalization, community and the moral economy of the farm crisis. Fernwood, Halifax
Luxton M (2007) Family responsibilities: the politics of love and care. In: Lucas LE (ed) Unpacking globalization: markets, gender, and work. Lexington, Lanham/Boulder/New York/Toronto/Plymouth, pp 131–144
Martz DK (2006) Canadian farm women and their families: restructuring, work, and decision making. (Unpublished doctoral dissertation), University of Saskatchewan, Saskatoon
McBride S (2001) Paradigm shift: globalization and the Canadian state. Fernwood Publishing, Halifax
Milne W (2005) Changing climate, uncertain future: considering rural women in climate change policies and strategies. Can Woman Stud 24(4):49–54

Norberg-Hodge H, Merrifield T, Gorelick S (2002) Bringing the food economy home: local alternatives to global agribusiness. Fernwood, Halifax

Novek J (2003) Intensive hog farming in Manitoba: transnational treadmills and local conflicts. Can Rev Sociol Anthropol 40(1):3–26

Orr DE, Shen Y (2006) World pig production: opportunity or threat?. Paper presented at the Indiana Farm Bureau's Midwest Swine Nutrition Conference, Indiana, 2006. http://www.thepigsite.com/articles/2/ai-genetics-reproduction/1857/world-pig-production-opportunity-or-threat

Panitch L, Gindin S (2009) The current crisis: a socialist perspective. Stud Pol Econ 83:7–31

Petras J, Veltmeyer H (2001) Globalization unmasked: imperialism in the 21st century. Fernwood, Halifax

Phillips EJ (1998) The social and cultural construction of farming practice: 'good' farming in two New South Wales communities. (Unpublished doctoral dissertation), Charles Sturt University, Australia

Public Safety Canada (n.d.) Canadian disaster database. Retrieved from http://cdd.publicsafety.gc.ca/srchpg-eng.aspx?dynamic=false.

Qualman D (2001) Corporate hog farming: the view from the family farm. In: Epp R, Whitson D (eds) Writing off the rural west: globalization, governments, and the transformation of rural communities. University of Alberta Press, Edmonton, pp 21–38

Sauchyn D, Kulshreshtha S (2008) Prairies. In: Lemmen D, Warren F, Bush E, Lacroix J (eds) From impacts to adaptation: Canada in a changing climate 2007. Government of Canada, Ottawa, pp 275–328

Shields J, Evans BM (1998) Shrinking the state: globalization and public administration "reform". Fernwood, Halifax

Sierra Club, Holbrook-White S (2000) NAFTA transportation corridors: approaches to assessing environmental impacts and alternatives. In: Proceedings from the North American symposium on understanding the linkages between trade and environment, Washington, DC. Accessed at http://www.cce.cec.org/programs_projects/trade_environ_econ/pdfs/sierra.pdf

Statistics Canada (2006a) The financial picture of farms in Canada. 2006 Census of Agriculture Highlights and Analyses. Statistics Canada, Ottawa. Retrieved from http://www.statcan.gc.ca/ca-ra2006/articles/finpicture-portrait-eng.htm

Statistics Canada (2006b) Snapshot of Canadian agriculture. 2006 Census of Agriculture Highlights and Analyses. Statistics Canada, Ottawa. Retrieved from http://www.statcan.gc.ca/ca-ra2006/articles/snapshot-portrait-eng.htm

Statistics Canada, Statistics and Data Development Unit (n.d. a) Table 1: exports of pork and live hogs, Alberta and Canada. Retrieved from Alberta agriculture at http://www1.agric.gov.ab.ca/$department/deptdocs.nsf/all/sdd11840/$FILE/table 1.pdf

Statistics Canada, Statistics and Data Development Unit (n.d. b) Table 15: Canadian exports of live hogs by province. Retrieved from Alberta agriculture at http://www1.agric.gov.ab.ca/$department/deptdocs.nsf/all/sdd11840/$FILE/table 15.pdf

Statistics Canada, Statistics and Data Development Unit (n.d. c) Table 23: Canadian exports of live hogs by country. Retrieved from Alberta agriculture at http://www1.agric.gov.ab.ca/$department/deptdocs.nsf/all/sdd11840/$FILE/table 22-23.pdf

Stirling B (2001) Work, knowledge, and the direction of farm life. In: Epp R, Whitson D (eds) Writing off the rural west: globalization, governments, and the transformation of rural communities. University of Alberta Press, Edmonton, pp 247–261

Tait F (2003) Pork, politics, and power. In: Ervin AM, Holtslander C, Qualman D, Sawa R (eds) Beyond factory farming: corporate hog barns and the threat to public health, the environment, and rural communities. Canadian Centre for Policy Alternatives, Saskatoon, pp 39–58

Thu K (2003) Industrial agriculture, democracy, and the future. In: Ervin AM, Holtslander C, Qualman D, Sawa R (eds) Beyond factory farming: corporate hog barns and the threat to public health, the environment, and rural communities. Canadian Centre for Policy Alternatives, Saskatoon, pp 9–28

Whatmore S (1991) Farming women: gender, work and family enterprise. Macmillan, Basingstoke

Wiebe N (1995) Farm women: cultivating hope and sowing change. In: Burt S, Code L (eds) Changing methods: feminists transforming practice. Broadview Press, Peterborough, pp 137–163

World Trade Organization (1999) Trade and environment, vol 4, Special Studies. WTO, Geneva

Young M (2005) Efficiency of pork production: a USA, Canada, and Ireland Comparison. Allen D. Leman Swine Conference, University of Minnesota, pp. 124

Chapter 9
Renegotiating Gender as Farming Families Manage Agricultural and Rural Restructuring in the Mallee

Josephine Clarke

Abstract This chapter explores climate change related farming and structural adjustment through a gender lens. It considers the political and public policy contexts associated with the gendered social impacts of climate change and draws on current research with women and men in the Mallee region of Victoria about the impacts of climate change. The chapter concludes that a gender analysis of farming adjustment policy makes visible the importance of understanding shifting gender relations as women and men manage agricultural and rural restructuring in a context of change that includes climate change.

Keywords Gender • Farming families • Rural restructuring • Agricultural policy • Gender relations

Introduction

Farming in Australia continues to be transformed for many reasons including climatic, environmental, economic, governance (policy), technological and social. Consequently changes are resulting in agricultural as well as rural restructuring (Pritchard and McManus 2000; Barr and Karunaratne 2002; Beer et al. 2003; Barr 2005). The research informing this chapter considers the experiences and outcomes for women and men who are currently farming as well as those who have exited farming. These experiences and changing gender relations are evolving in the context of climate change, future scenarios, and uncertainty.

Agricultural production, farm structures and rural communities in Australia, have undergone significant adjustment since World War Two and in particular,

J. Clarke (✉)
Gender, Leadership and Social Sustainability Research Unit, Department of Social Work, Monash University, Caulfield, VIC, Australia
e-mail: josephine.clarke@monash.edu

M. Alston and K. Whittenbury (eds.), *Research, Action and Policy: Addressing the Gendered Impacts of Climate Change*, DOI 10.1007/978-94-007-5518-5_9,

since the 1970s. The after-war period until the 1970s is described as a profitable time for Australian agricultural production (as well as other industries) and rural communities (Lawrence and Vanclay 1994; Beer et al. 2003). An economic and political culture developing since the 1970s supports privatisation, deregulation, and a reduction in trade tariffs and protectionism for Australian agricultural produce. This period is characterised by industry and rural restructuring. 'The term restructuring is commonly used to characterize shifts in the economic trajectories and political strategies of advanced nations following the end of the postwar boom in the 1970s' (Tonts 2000). Since this time the predominantly male agricultural workforce has declined in numbers, is older, paid lower wages, and increasingly casualised (Barr 2009). Women's unpaid on-farm labour has been a feature of family farming (Alston 1995) as well as off-farm work to support farm viability in times of drought (Alston 2006). These changes have occurred at the same time overall agricultural production has increased, assisted by technological developments including new machinery and yield gains achieved through genetics, fertilizers and herbicides (Productivity Commission 2005).

The Mallee is a region in northwest Victoria, Australia, with substantial agricultural and horticultural production. The Mallee is a low rainfall region. Many localities in the Mallee typically receive 300–350 mm of rain on average and the region has become drier and hotter over recent decades and experienced a number of droughts. The region includes a large area of dryland broadacre farming (mixed crops and sheep) as well as irrigated food production (nuts, dried fruit, table and wine grapes, citrus) along the Murray River.[1] Irrigation farming is dependent on the Murray-Darling Basin river systems which has been identified as a 'hot spot' with respect to climate change (Hennessy et al. 2007). The Mallee region is expected to continue to become drier and warmer due to climate change (Department of Primary Industries 2009). It is possible that yield productions may decrease and production sites shift spatially (Department of Primary Industries 2008).

Family farming still dominates Australian agricultural production in terms of the number of farms (Productivity Commission 2005) although there has been a decline in the number of family farms. Small farms dominate in Victoria and many are located in the Mallee region with a higher concentration along the Murray River due to horticultural production (Barr and Karunaratne 2002, p.5). However, with respect to broadacre farms, this type of farm has generally increased in size and farmers in these regions have been challenged to 'get big or get out' due to changes in agricultural productivity and terms of trade that put pressure on farms to enlarge in order to remain financially viable (Barr 2009).

Similar to other rural areas in Victoria (and elsewhere in Australia), there has been a demographic shift resulting in population decline, with in-migration into larger regional centres in the Mallee region (Barr 2005; McKenzie and Frieden 2010). This includes a decline in the number of young people in rural areas, as

[1] Irrigated farming areas in the northern Mallee area are also collectively known as the Sunraysia region.

well as out-migration due to farm exits (Alston 2004; Geldens 2007). Demographic changes and the reduction of the number of farms have resulted in ensuing changes for small towns. Many Mallee towns in the dryland cropping areas no longer have schools, shops or other services located in the town.

Consequently, this research recognises that farming women and men are experiencing and managing an array of changes in agricultural production in a social and governance context that is informed by future scenarios and uncertainty. This context may include a number of issues at any given time such as farm succession, aspirations for family members and livelihood strategies. Increasingly climate change is a major component of agricultural and rural restructuring.

Climate change refers to those changes in long-term weather patterns that indicate increases in greenhouse gases and a rise in global surface temperatures. Further, these climate changes are attributed to anthropogenic activities (Department of Climate Change and Energy Efficiency 2011). Climate change science is not static: for example, in 2010 temperatures were predicted to increase by approximately 2°.[2] In 2011 there has been further information and science describing this current decade as the 'critical decade' for mitigation of the projected impacts of climate change in order to avoid 4° scenarios – that is an increase in surface temperatures by 4° – by the end of the century, 2100 (Meinshausen 2011). In turn there are governance and public policy discourses that refer to climate change science and consequently have social impacts. Multiple aspects to 'climate change' immediately converge: climate change as a measurable climatic trend as defined by scientific discourses exists in combination with local experience; public policy is developed in relation to climate change science and information relating to agricultural and rural restructuring resource distribution.

The Political and Policy Context to Climate Change in Australia

Bacchi's (2009) work provides a useful framework to analyse public policy. Drawing on a Foucault her analysis involves six aspects that include:

- a critical analysis of the policy 'problem' being represented;
- an examination of the underlying assumptions that establish the policy-problem;
- an understanding of the discursive history informing the policy-problem;
- a review of the 'silences' and understanding what is not being stated in the policy with a view to considering alternative understandings of the policy-problem;
- a questioning of the powerful effects of the policy; and

[2]A rise in temperatures of 2° by 2070 is a prediction for the Mallee region (with seasonal and locality variations as well as scenario variations). See http://www.climatechange.vic.gov.au/__data/assets/pdf_file/0012/73101/Mallee_WEB.pdf. Temperature rises are predicted to vary across the globe.

- an understanding of where the policy 'problem' is sited in terms of where is it located and where/whom it is dominating as well as considering how the policy can be challenged and alternatives developed.

Bacci's policy-problem approach assists a critical analysis of Victorian agricultural restructuring policy. This approach draws attention to the significance of public policy documents as a source of data that draw attention to government defined priority issues impacting upon social structures and relations including family farming.

The Victorian Government's Future Farming (Department of Primary Industries 2008) policy strategy, *Future Farming: Productive, Competitive and Sustainable* was released in 2008 and details the Victorian Government's policies and strategies to support farming. A paramount feature of this policy is the emphasis on the need for farmers to manage changes in farming and manage an array of challenges that include commodity price uncertainty, climate change, technological supports, and proposed changes to farm business structure. The policy document specifically considers the future of family farming, for example:

> The challenges taking place in farming have particular implications for small to medium family farm businesses. While some may be supported by off-farm wages, others may need to expand, restructure their operations or change long-standing practices...
>
> The new era in farming is being driven by economic, climate and market forces largely outside our control. Attempting to slow down or hold back the rate of change will generate pressure for even more rapid and painful adjustment later. (7)

Therefore, discourses of family farming now include a government policy expectation that family farming will continue to decline (Department of Primary Industries 2008). This expectation informs a policy 'action' focused on providing supports to assist farm health and decision-making including rural financial counseling support for farmers and families who decide to leave farming. An increased emphasis on decision-making is evident in adaptation strategies and there exists a tension in the 'choice' inferred when often farming women and men may feel forced from farming, or displaced (Bryant 1989).

This policy specifically targets an action focused on opportunities in farming for young people. Within the policy action description it is stated that:

While the traditional farming family as owner and operator will remain the most common approach for the foreseeable future, changing farm business models will provide new opportunities. Increasingly, professional farm managers will manage other people's land. Operations that require capital-intensive or specialist machinery, such as harvesting, will be carried out by specialist contractors.

In many cases, these specialist providers will not live on a farm. They will base themselves in local towns and provincial centres, where they can access community and social services. (17)

This articulation of the rise of a restructured farm is a move away from the family farm owner-operator model, and is also an imagining of rural restructuring and changing demographics of towns and regional centres. Implications also include changing social relations and agricultural and farming policy. It presents a future scenario for family farm businesses yet does not explicitly reference possible changes to roles and responsibilities for women and men involved in farming. Considering the policy cited above, it is important to consider *who* is imagined to be working off-farm in a *supporting* role, for example. It is well documented that women have increasingly contributed to farm viability by working off-farm (Alston 2006). In this policy document the farm as a site of production remains the reference point and agricultural production becomes sustainable only if farm restructuring follows a government nominated trajectory. However, the preferred restructured farm as a site of production is interacting with 'forces' the government does not exert much control over i.e. economic and market domains and climate uncertainty.

This is a policy moment where climate change and terms of trade merge in an uncertain future scenario and yet at the same time a policy-prescribed farming business is prescribed that will successfully manage a coalescence of changes. If, given the scenario the policy presents, farm businesses will involve less family-labour in the future, what is the future for women in farming and what are the emerging opportunities as a range of trade, environmental, climatic, and technological changes facilitate agricultural and farm restructuring?

A plethora of research has documented the patrilineal and patriarchal nature of farming with women often providing on-farm labour as well as working in the home (Sachs 1983; Whatmore 1991; Alston 1995). This policy asserts the separation of family from farm business and in turn this demands a revised relationship by the 'farmer' with the land. The restructuring nominated also demands revised gender relationships, roles and responsibilities yet these are not mentioned in the policy document. It is young people's careers in farming that are emphasized while no mention is made of opportunities for women or any other categorical social group.

The policy offers a scenario for agricultural restructuring whereby the instability of the family farm is challenged by multiple external forces that are non-corporeal, that is, economic, market and climate changes are unable to be influenced by the government or indeed the farmer or farming family as they are not situated in a time or space. The forces are nowhere even though they are predicted to impact upon lives and livelihoods. Here the policy defines the role of government and by asserting that there are forces largely beyond its control, it affirms a strategic intent to support a type of farm restructuring. Farm exits are not overtly named but anticipated in this uncertain future. There is a focus on engagement with the farming family to enable competitive farming, defined largely on higher production levels and a revised farm business structure. Climate change becomes an external (and uncertain) influence driving a specific policy based on a presumed restructured farm enterprise. The policy conversation precludes a focus on social equity issues involved in restructuring. Thus climate change *impacts upon* agricultural production in a specific way in this policy context. In this scenario climate change becomes a destabilising force and cuts off the potential for policy actions to support social engagement. In other words, the policy is promoting a move away from family farming to a very specific farm structure without any consideration of the social equity issues arising from farm restructuring. In renaming terms of trade – now including climate change – as beyond government influence there is a danger in this policy framework of avoiding socially responsible agricultural restructuring and hence food production.

'Climate change', with its social, historical and political context, informs adaptation and is a powerful discourse. In informing policy development, it increasingly prompts rural and agricultural structural readjustment initiatives. Given the rise of multiple uncertainties, it is important to understand how this influences gender negotiations.

Climate change is anticipated to compound issues of women's lack of access to resources and participation in decision-making processes (Terry 2009). Importantly, the way 'future scenarios' are being constructed and challenged (Milne 2005) without due consideration of gender is disturbing. Policy attempts to direct farming currently do not include women or promote equal participation in conversations about 'the future'. Structural adjustments can be diverse especially if opportunities are provided to include more women in agriculture and challenge current farm inheritance and farming practices. In this context, agricultural policy can address social and gender equity issues.

Researching the Social Impacts of Climate Change

Until recently many parts of south-eastern Australia were experiencing a major drought that had lasted for over a decade. This impacted upon dryland farmers who rely on rainfall, and irrigators who experienced reduced water allocations and the introduction of water trading. In early 2011 there was flooding in a major regional centre that supports irrigation farming and substantial rainfall in the Mallee region. Initial interviews were undertaken with key informants 1 month after this severe rainfall event.

> What the rain has done has, it's taken people that would have probably would have ended up out of farming albeit in 12–18 months time, its accelerated their exit to the point that now it is dramatic and there's no think time, its bang it's going to happen and that in itself is, people haven't got the time to prepare mentally for that themselves ... we don't know what the impact of that is going to be (*Key Informant*)
>
> I think there was expectation that we are going to get out of drought... and things will be better ... [we] came out of drought but the door slammed shut in our face. (*Key Informant*)

Key informants note the pressure of managing the impacts of an extreme rainfall event compounding existing stressors on farmers who have been managing drought and changes in water allocations. Further, key informants identify that this weather event will prompt farm exits.

In a preliminary interview 'Chris', a dryland woman farmer, raises the issue of climate change predictions, and asserts her local knowledge of the weather and climate cycles.

Chris: And you see this is the weather, you're dependent. And this is what I can't understand about how these people scientists think they can model what is going to happen because the modelling is only as good as what they put in and

Jo: ... climate change predictions?

Chris: yes that's right, and predictions, they can't, well even following the weather report on there, they can't predict five days out. How can they predict – that's the way I feel anyhow. And because we live with the weather we know how unpredictable it can be. So I just don't put a lot of credence in what they say.

Jo: Because I'm aware this is the driest part of state and it's predicted to get drier.

Chris: Yes. But that – if that happens it doesn't worry me particularly because, they adapt. We adapt our machinery, we're not um growing crops like we did when I first came on the farm. We've adapted, we are much more environmentally conscious, we used to plough the crop, plough the ground up, you'd do it seven or eight times to get it really fine then the wind would come and blow them and that's how you'd get all these dust storms. Now we don't break the soil up at all ...

Chris articulates her experiences of adaptation and faith that adaptation will continue. So 'climate change' is not a static or neatly-bounded thing/experience/possibility. It is science plus many other discourses – local and policy for example – that structure farm decision-making.

A challenge for this research is that by accepting climate change predictions for the Mallee region, and interpreting those predictions as part of agricultural and rural restructuring, we could risk adding climate change predictions to what Herbert-Cheshire (2003) describes as a'rural crisis discourse'. Climate change predictions potentially add to a litany of difficulties for agriculturally-dominated rural regions – low commodity prices, changing conditions of trade, changing land use, an ageing rural population, loss of service infrastructure as well as extreme weather events. As Barnett (2006) has emphasized, situating climate change alone as impacting upon people risks ignoring other existing factors at work contributing to social vulnerability that in turn are compounded by climate change.

Conclusion: Gender Analysis, Structural Readjustment and Working Towards Gender Equity

In this research I challenge a discourse dominating rural policy in Australia – a masculinist, productivist construction. This focus is implicit in the Victorian Government's Future Farming policy (Department of Primary Industries 2008) and ignores dynamic social relations. My research with farming family households interrogates the way discursive constructions infusing policy affect, and are affected by, a gender order that renders farming women invisible (Connell 2005).

In researching the gender order – critical to understanding how women and men are coping with and adapting to multiple changes in farming – I am working with women and men who are currently farming as well as those who have left farming. Again, this is a strategy to acknowledge the diversity of experiences in managing agricultural and rural restructuring and challenge the farming/exit farming binary opposition that prevails in policy. It is important to acknowledge increasingly diverse rural livelihoods and pursue research that seeks to understand the diversity of experiences that constitute the 'outcomes' of agricultural restructuring. In this way gender focused research can assist in working towards sustainable social and gender relations in agriculture and inform policy in the context of adapting to climate change.

References

Alston M (1995) Women on the land: the hidden heart of rural Australia. UNSW Press, Kensington

Alston M (2004) 'You don't want to be a check-out chick all your life': the out-migration of young people from Australia's small rural towns. Aust J Soc Issues 39(3):299–313

Alston M (2006) 'I'd like to just walk out of here': Australian women's experience of drought. Sociol Rural 45(2):154–170

Bacchi C (2009) Analysing policy: what's the problem represented to be? Pearson, French Forest

Barnett J (2006) Climate change, insecurity, and injustice. In: Adger WN, Paavola J, Huq S, Mace MJ (eds) Fairness in adaptation to climate change. MIT Press, Cambridge

Barr N (2005) Understanding rural Victoria. Department of Primary Industries, Bendigo

Barr N (2009) The house on the hill: the transformation of Australia's farming communities. Land & Water Australia, Canberra

Barr N, Karunaratne K (2002) Victoria's small farms, CLPR Research Report 89. Department of Natural Resources and Environment, Bendigo

Beer A, Maude A et al (2003) Developing Australia's regions: theory and practice. UNSW Press, Sydney

Bryant L (1989) The Resettlement Process of Displaced Farm Families: A Study of 12 Families From Eyre Peninsula, South Australia, Technical Paper No. 27. Department of Agriculture, South Australia, Adelaide

Connell R (2005) Change among the gatekeepers: men, masculinities, and gender equality in the global arena. Signs J Women Cult Soc 30(3):1801–1825

Department of Climate Change and Energy Efficiency (2011) Understanding climate change. Retrieved 29 Jan 2012 from http://www.climatechange.gov.au/en/climate-change/understanding-climate-change.aspx

Department of Primary Industries (2008) Future farming: productive, competitive and sustainable. Victorian Government/Department of Primary Industries, Melbourne

Department of Primary Industries (2009) Primary production landscapes of Victoria – Northwest Victoria. Retrieved 28 Mar 2010 from http://www.dse.vic.gov.au/dpi/vro/vrosite.nsf/pages/primary_prod_landscapes_nw_vic

Geldens P (2007) Out-migration: young Victorians and the family farm. People Place 15(1):80–87

Hennessy K, Fitzharris B et al (2007) Australia and New Zealand. In: Parry ML, Canziani OF, Palutikof JP, van der Linden PJ, Hanson CE (eds) Climate change 2007: impacts, adaptation and vulnerability. Contribution of working group II to the fourth assessment report of the intergovernmental panel on climate change. Cambridge University Press, Cambridge, pp 507–540

Herbert-Cheshire L (2003) Translating policy: power and action in Australia's country towns. Sociol Rural 43(4):454–473

Lawrence G, Vanclay F (1994) Agricultural change in the semiperiphery: the Murray-darling basin, Australia. In: McMichael P (ed) The global restructuring of agro-food systems. Cornell University Press, Ithaca, pp 76–103

McKenzie F, Frieden J (2010) Regional Victoria: trends and prospects. Victorian Government/Department of Planning and Community Development, Melbourne

Meinshausen M (2011) Climate risks, probabilities, and projections conference paper. Four degrees or more? Australia in a hot world conference, University of Melbourne, Parkville, 12–14th July 2011

Milne W (2005) Changing climate, uncertain future: considering rural women in climate change policies and strategies. Can Woman Stud/Les Cah De La Femme 24(4):49–54

Pritchard B, McManus P (eds) (2000) Land of discontent: the dynamics of change in rural and regional Australia. UNSW Press, Sydney

Productivity Commission (2005) Trends in Australian agriculture. Research Paper. Commonwealth of Australia, Canberra

Sachs C (1983) The invisible farmers: women in agricultural production. Rowman & Allanheld, Totowa

Terry G (2009) No climate justice without gender justice: an overview of the issues. Gend Dev 17(1):5–18

Tonts M (2000) The restructuring of Australia's rural communities. In: Pritchard B, McManus P (eds) Land of discontent: the dynamics of change in rural and regional Australia. UNSW Press, Sydney

Whatmore S (1991) Farming women: gender, work and family enterprise. Macmillan, Basingstoke

Part IV
Action and Strategies to Address Gender and Climate Change

Chapter 10
Gendered Access to Green Power: Motivations and Barriers for Changing the Energy Provider

Gotelind Alber

Abstract Sustainable energy supply is a prerequisite for sustainable development. Through their energy consumption customers influence the extent of energy use, and by choosing their energy provider they can additionally influence how electricity is produced. This article summarises the finding of a research project on women and green power funded by the German Ministry for the Environment. A starting point of the project was existing evidence that women reject nuclear energy more strongly than men. The objectives of the project were to analyse the motivation of women to switch to green power and the barriers they experience, to test intervention strategies and to provide recommendations for further action to promote green power among women. The data related to motivations and obstacles to change the energy provider was generated by an online questionnaire with more than 1,000 respondents, and by qualitative interviews reflecting the heterogeneity of female lifestyles. Furthermore, interventions to initiate the shift to renewable, nuclear free energy were developed, tested and evaluated. The interventions were based on the assumption, that women trust more in face to face communication and in experiences communicated by people they know, than in anonymous, seemingly objective information and estimations which are preferred by men.

Partners in the interdisciplinary research project were LIFE e.V. education | environment | equality, Berlin, the Free University Berlin, and Sustainable Energy and Climate Policy, Berlin.

Keywords Green power • Energy provider • Gender • Interventions • Germany

G. Alber (✉)
Sustainable Energy and Climate Policy, Berlin, Germany
e-mail: go@gotelind-alber.eu

M. Alston and K. Whittenbury (eds.), *Research, Action and Policy: Addressing the Gendered Impacts of Climate Change*, DOI 10.1007/978-94-007-5518-5_10,

Introduction

The expansion of renewable energies in Germany is a success story. During recent years, a major increase of the share of renewable energy was achieved, with annual growth rates of up to 75%, in particular in the electricity sector. Germany has large production capacities for renewable energy. The number of jobs created in this sector outmatches those in the conventional energy sector, and the sectors' value added in the year 2010 was almost 38 billion Euro.

However, there is still much to be done. The phase-out of nuclear has been decided by the government but there is no guarantee that this decision will be implemented. After all, there had already been a "nuclear consensus" in the year 2000 to abandon nuclear energy. Even though this was based on an agreement with the utilities, the current government decided in 2009 to revoke this consensus, extending the life of the nuclear reactors by more than 10 years. After the nuclear accident in Fukushima, another U-turn was taken, and the government announced Germany will abandon nuclear power completely by 2022. However, the nuclear lobby is still working hard in order to achieve a third U-turn back to nuclear which provided 18% of Germany's electricity production in the year 2011. If a complete phase-out of nuclear is to become a reality, renewable energies which, today, cover some 20% of electricity demand, need to be boosted even the more.

Despite the favourable conditions for electricity generation from renewable energy in Germany, the purchase of green power still makes sense if the provider guarantees that new power plants based on renewable energy sources will be built with the additional proceeds. Ultimately, more capital will be available for the expansion of renewable energy, so that with an unchanged level of electricity consumption, conventionally generated power can be displaced. Moreover, a signal will be sent to conventional energy providers that consumers desire a restructuring and democratisation of energy supply.

Women in Germany endorse strong climate protection policies more clearly than men (European Commission 2009; Forsa 2007), and reject nuclear energy more strongly. Seventy-two percent of German women reject nuclear power, compared to 60% of men, 42% of women prefer an immediate phase-out of nuclear, compared to 34% of men, and 55% of women consider themselves as green electricity aficionadas, compared to 44% of men (Borgstedt et al. 2010; Marktforschung.de 2011). In terms of women's consumption choices, criteria relating to ethics and fairness play a decisive role (Schultz and Stiess 2009). These findings suggest that women are very likely to be amenable to a switch to green energy. However, this is not directly reflected in reality, and the number of households that switch to a green energy provider or at least a green energy tariff is still low. According to data obtained through a recent survey about green power (Energie and Management 2011) which takes into account most providers, three million households in Germany were purchasing green energy at the end of 2010. Even after a strong increase of green power customers as a response to nuclear disaster in Fukushima, the proportion of households purchasing green energy remains at only 7.5%.

In particular studies on environmental consciousness (Kuckartz and Rheingans-Heintze 2004; Kuckartz et al. 2006), when people who had not switched to green power were asked about their reasons for this, they cited lack of information and high costs, as well as the effort involved in making the switch. No differences between genders were observed in these responses. Other studies, such as the one carried out by Ecologic on the effects of the environmental tax reform in Germany (Umweltbundesamt 2004), show clear gender differences with respect to the impact of the higher electricity prices on energy-saving behaviour. These higher prices prompted almost half of the women (49%) but only 36% of the men to save energy. From this, it can be presumed that the higher price of green power is likely to play a larger role for women than for men.

The responsibility for electricity is distributed in different ways in couple households. Even though research about consumption (Johnsson-Latham 2006, 2007; Weller 2008; Weller et al. 2002) has shown that women make a disproportionately high share of everyday consumption decisions in private households, women have not been shown to bear any higher level of responsibility in the area of energy. Then again, both money and time, which can be restrictive in terms of switching energy providers, are tied up with gender roles – in particular, with the gender-specific division of labour and the frequent 'double burden' borne by women.

Several studies (Birzle-Harder and Götz 2001; Iyengar and Lepper 2000; Sunderer 2006) indicate that many people who are surveyed know very little about green energy, are uncertain about how the system functions and would like simpler access to information. The fact that information and materials cannot sufficiently answer unresolved questions is also criticised. The problem of information provision is made worse as a result of the intricacy of the issue (where green energy is sourced, how it is produced, how it is distributed in the grid ...) and the diversity of offers from energy providers. These difficulties are heightened by a lack of credibility about the environmental benefits of green power. Consumers can only ascertain the environmental advantages on the basis of data that they cannot review, especially since there is no official definition of green power. Building confidence and transparency therefore plays a significant role in motivating people to switch to green energy, and this brings corporate policies in general into the picture.

Sunderer (2006) comes to the conclusion that green energy as a product is subject to active and intensive information and communications policies. Without having the issue brought to their attention, consumers would rarely think about switching of their own accord – people do not often seek out information independently. Because green power is an abstract product that requires explanation, and since all studies on the subject have pointed out problems relating to credibility, information should not simply be placed in impersonal communication materials such as brochures or posters, but should strengthen the direct dialogue with consumers.

Large amounts of information are available about electricity providers and tariffs. However, such information is biased according to the different interests of the information providers and requires pre-existing technical knowledge for people to be able to classify and, in particular, assess the options. The German electricity market – including the green power market – is confusing and unclear; a multitude of tariffs and labels obscure its essential aspects from view. This can be seen, for instance, in the fact that a considerably higher proportion of consumers (15%) believe that they would purchase green power, but do not actually do so (Immowelt 2010). Furthermore, with respect to information provision, the fact that women may prefer different information channels to men is ignored (Kuckartz et al. 2006), and therefore women as a target group are often not reached. According to market research results (for instance Barletta 2003; Underhill 2000), women are more persuaded by knowledge they gain through experience – that is, knowledge imparted through direct exchange, for instance from friends, acquaintances and colleagues – rather than printed information or factual knowledge on internet sites.

With this background in mind, the research project was designed to shed light on how decisions about electricity providers are made, who is involved in the decisions, and how and/or through what means people inform themselves. On the basis of these findings, intervention strategies aimed at motivating women to switch to green energy were developed and tested.

The following hypotheses guided the research:

1. Based on their anti-nuclear stance, women are amenable to the switch to green energy ("phase out nuclear ourselves")
2. Women prefer different information channels to men, which is why information about options for switching electricity does not reach them.
3. Because of the complexity of the electricity market and the different labelling involved, the assessment of green energy options is difficult and prevents people from switching to green energy.
4. Energy, and electricity in particular, is a topic that has male connotations, which is reflected in the responsibility for decisions on energy in the household and also has an impact with respect to switching energy providers.
5. The gender pay gap, in combination with the higher costs of green energy, prevents female-led households from switching to a green energy provider.

The Survey

The data related to motivations and obstacles to change the energy provider was generated by an online questionnaire with more than 1,000 respondents, and by qualitative interviews reflecting the heterogeneity of female lifestyles. The latter were supposed to provide a better understanding of how the decisions related to energy consumption are made, who is involved in this process and of gender as a decisive criteria for the decision-making process.

The online questionnaire was aimed at both women and men and incorporated a number of topics: environmental attitudes, switching of energy providers (motives and constraints), communications channels, decision-making structures and division of labour in the household. In addition, particular socio-demographic characteristics were requested. Over the survey period, a total of 1,526 participants filled out the online questionnaire, and 1,089 fully completed questionnaires were able to be used.

On average, the participants were 35.6 years old. The number of women in the random sample amounted to 55.2%. The education level within the random sample can be regarded as high, with 56.4% of respondents stating that they had a degree from a university or technical college. With respect to professions, students made up the largest group (33.6%). More than two thirds of the participants stated that they lived in a city (70.4%). The average net monthly income was between 1,500 and 2,000 Euro.

Main Findings of the Survey

As in other studies (Kuckartz et al. 2006), our online survey confirmed that women reject nuclear energy more strongly than men. Overall, environmental protection and the rejection of nuclear energy are the strongest reasons for switching to green energy for both women and men. More so than men, women – in their role as consumers – want to have an influence on societal development through switching electricity providers (see Fig. 10.1).

The initial impetus for both women and men to switch to green power came mainly from themselves, but for women less often, while for these, the feedback they received from acquaintances about their energy switch was more important than for men (see Fig. 10.2). Whilst the most important factor for both men and women was that renewable energy is genuinely promoted through the purchase of green energy, women place greater importance on the conditions of the switch to green power, for example its easiness.

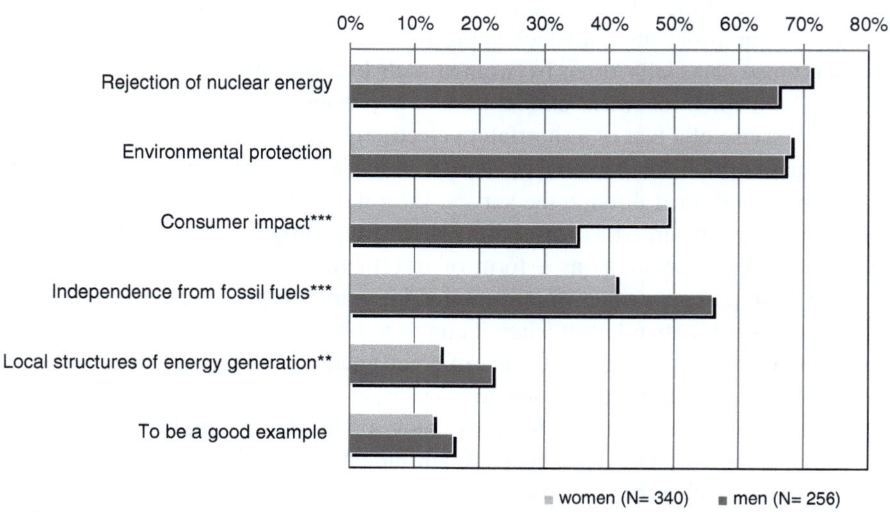

Fig. 10.1 Motives of purchasers of green power

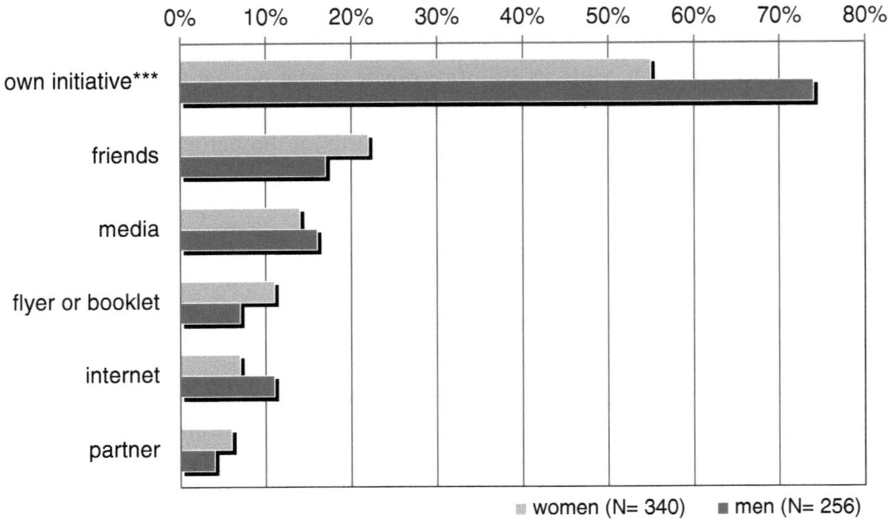

Fig. 10.2 Impetus for the switch to green power

Switching to green energy is a process – generally, people do not make the change immediately. Forty-eight percent of green power purchasers considered the decision "for some time", and for 14% it took "a long time" until they finally switched. Complacency or execution gap is cited most often by women and men as a constraint. Costs are cited significantly more often by men as a constraint, even though the men in our sample earned, on average, more than the women. For women, information about costs is also important, but higher costs act less often as a constraint. In contrast, the complexity and inaccessibility of the electricity market is regarded by many women as one area in which appropriate information is lacking. This reason is mentioned especially often by women with low educational qualifications and elderly women. In contrast, almost no-one is prevented from switching to green energy because of a fear of a shortfall in electricity supply (see Fig. 10.3).

The assumption that personal communication is more important for women than for men is confirmed through the survey. Information from friends, acquaintances and relatives is regarded as trustworthy more frequently by women than men, as are personal conversations as a form of information transfer. Acquaintances are especially important sources of information for young women and female students. The internet is highly relevant when it comes to obtaining information about green power. On the internet, the homepages of green energy providers and environmental organisations, as well as price comparison sites, are particularly helpful. As sources of information, green energy providers enjoy a considerably higher level of trust than conventional energy providers. Environmental organisations and consumer associations are also seen as trustworthy, while federal and local governments play a subordinate role when people seek to obtain information on the topic of green energy (Figs. 10.4 and 10.5).

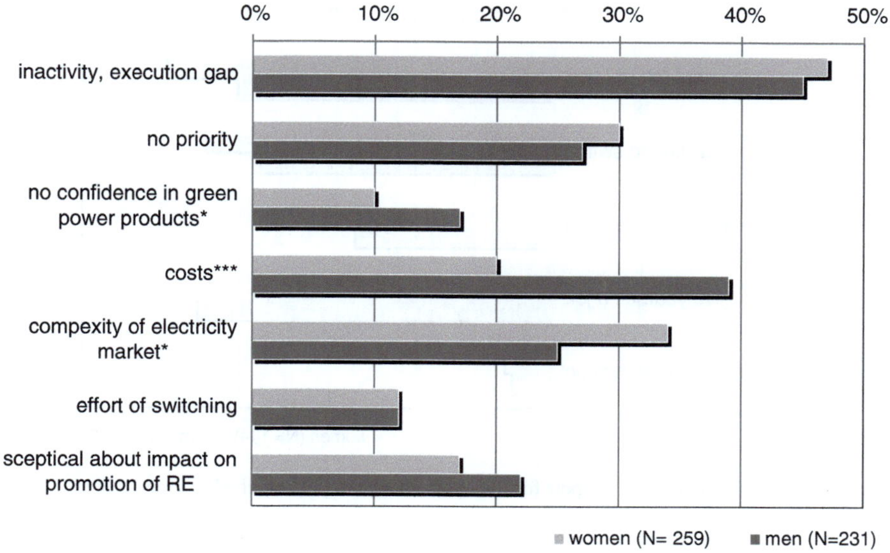

Fig. 10.3 Obstacles to switching to green power

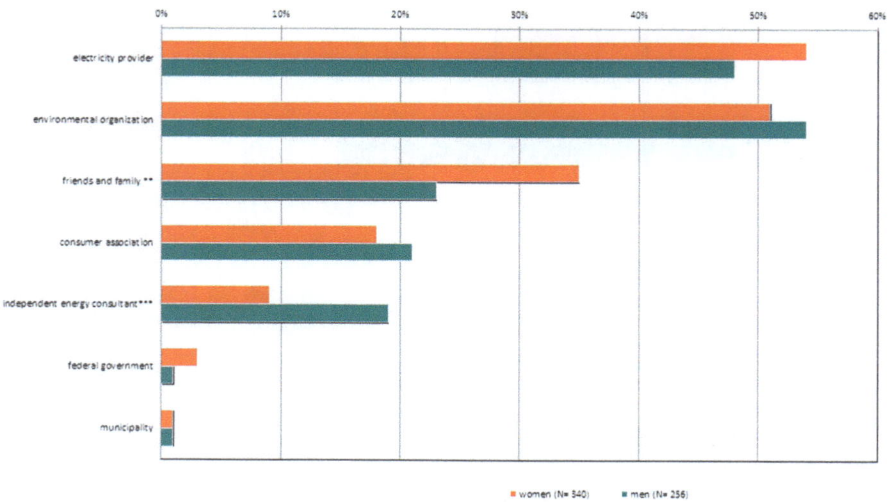

Fig. 10.4 Most important information resources for those who have already changed their electricity supplier

In more than half the cases surveyed, decisions about energy providers are made jointly. In couple-households it is more often the men than the women who make decisions about electricity supply. Overall, the decision about which energy provider to choose is seen to be relatively conflict free (Fig. 10.6).

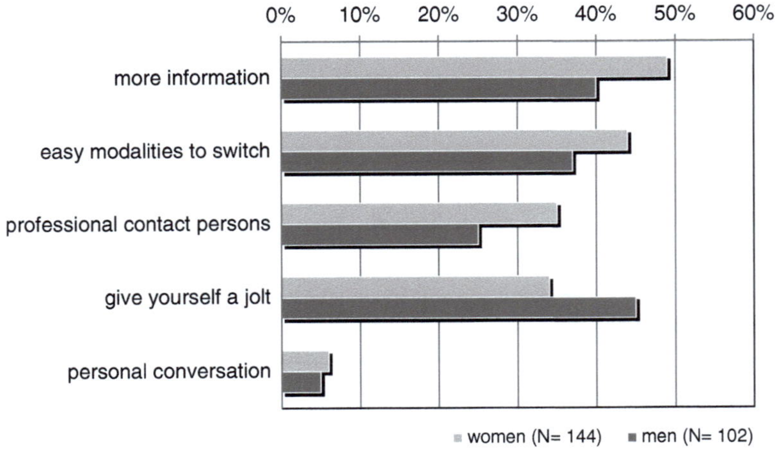

Fig. 10.5 Perceived need for support for those who have not yet switched to green power

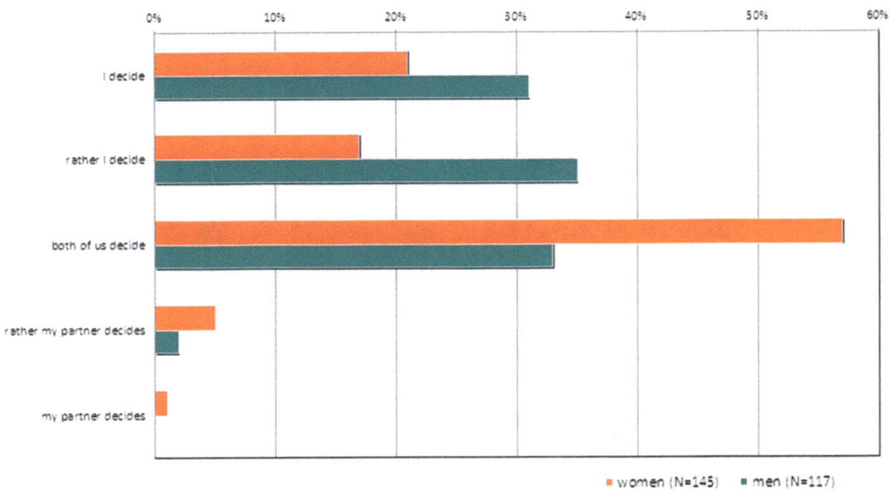

Fig. 10.6 Decision-making in the household

Qualitative interviews were held with 15 women of different German regions, educational backgrounds and income groups. Responses showed that the difficulty for environmentally conscious women is not making a decision for or against a switch to green energy. Rather, what is difficult is the actual implementation of the decision. Both the practical and the perceived effort of looking for a provider can delay the switch and/or prevent it from happening. When women experience similar constraints, differences can be observed that relate to particular living situations. Income and workload, as well as personal surroundings, play a role in determining how easy it is to switch to green energy and which particular constraints have the most impact.

It became clear through the interviews that the choice of an energy provider differs significantly from other consumption decisions. On the one hand, this relates to the abstract nature of electricity, which does not change its physical properties when it is 'green' electricity. On the other hand, the switch to green power is less about purchasing a sustainable product than entering a contract with a (sometimes) more sustainable company.

Green Power Training for Women

The aim of the training was to enable participants to overcome obstacles to switching their energy providers and to acquire the necessary knowledge and confidence to approach other people about switching to green power.

Therefore, both the practical aspects of switching to green electricity and the relevant background knowledge were included in the training in order to equip participants with points to convince people in their own households and other third parties. Furthermore, financial aspects were addressed to enable women to ascertain potential increases or reductions in costs and recognise savings opportunities. The training was provider-neutral, but did offer assistance in terms of assessing different providers.

A central element of the approach was the gender-sensitive design of the training sessions and the accompanying materials, in light of existing recommendations and guidelines (see, for example, Gindl et al. 2007). Given the topic of focus, a good starting point was to offer the training exclusively to women and work with female trainers. In this way, conditions were created that were more conducive to learning for most women – because, for instance, the rhetoric associated with technically orientated men could be avoided and it was easier to ask follow-up questions, even when it demonstrated a lack of technical knowledge on the part of participants (Deutsches Institut für Erwachsenenbildung 1994; Faulstich-Wieland et al. 2008; Gieseke 2001). In addition, the accompanying materials were prepared by female experts.

The content was prepared to be as practical as possible, without omitting technical matters, which were covered in only the necessary amount of detail. The training covered:

- the practical aspects of switching to green energy (formalities and modalities such as security of supply and reading of the electricity meter);
- selection of a green energy provider (basic information about the structure of the electricity industry in Germany; selection criteria such as additional benefits of green energy, company policies and services, concrete comparisons of typical providers, including costs, energy mix, source of energy, additional benefits and services; green power labels and renewable energy certificate system (RECS), fundamental aspects and operation of the German Renewable Energy Law) and

- household energy-saving options (significance of different uses of energy in the home, purchase of appliances and efficiency ratings, stand-by energy use and how to avoid it, lighting, including a comparison of CFL and LED lights, hidden consumers of energy, such as circulation pumps for heating and hot water, resources such as energy consumption calculators and top-rated lists).

Because of women's chronic shortage of time (see, for instance, Turner and Grieco 2000) – the training was designed as a short workshop of a maximum of 4 h. It was carried out in small groups with as many interactive elements as possible, so that any uncertainties could be dealt with in a targeted way and participants would be prepared for future situations and conversations. It was therefore part of the approach to create a relaxed atmosphere, allow questions at any time and deal directly with the needs of the participants.

Evaluation of the Training Measures

Fifty percent of the participants responded to a questionnaire which was sent to them 3–4 months after the training session. Forty-four percent of the respondents had already purchased green power prior to the training, and attended the training session to gather more background information in order to promote green power. Thirty-five percent had actually switched to green power, while 21% felt encouraged to do so, but had not yet implemented their decision. This confirmed the findings of the online survey that switching is a process that takes time.

The training sessions were regarded by almost all participants as being very important for their own switch in energy providers and their role as key communicators and information disseminators. Indeed, it became clear during the training sessions – and this was confirmed in the interviews – that the topic of green energy is very complicated and hard to penetrate for 'lay' people. From the start, this led participants to voice doubts about how even they as trainers can pass on this knowledge. The difficulties in this sense were not so much about the process of switching, but rather about deciding on a provider or a tariff, which then led to questions about the German renewable energy law, green power labels and the purchase of certificates (RECS). This shows that in relation to technical issues, it is particularly important to women to really penetrate a topic and understand it completely before they further impart their knowledge. In contrast, men tend to accept something as fact and pass it on with corresponding (technical) authority.

Nevertheless, to a large extent, the participants in the training sessions were themselves active as information disseminators and sought out their own ways to persuade others. Eighty-two percent of participants believed that personal communication is essential, while only 6% deemed a leaflet to be sufficient.

Ninety-four percent talked with friends and colleagues about green power. Moreover, 13% participated in other activities such as information stands and 6% arranged events such as "green power parties". Only 3% did not promote green power, as they felt they did not want to "proselytize". Most of the participants tried things out for themselves to determine what worked best, and were very successful in doing so. These approaches tended to consist of many small, understated steps, rather than events. Thus, the multiplier effect could only be measured during the first stage of dissemination. Forty-five percent of those who actually promoted green power knew that some of their friends changed the supplier (some persons on average per promoter). Forty percent felt that these would, in turn, promote green power. According to telephone interviews which were carried out with some of the respondents, the multiplier effect (or diffusion) does in fact take place at later stages, but is no longer able to be quantitatively ascertained.

Conclusions and Recommendations

In conclusion, we have found the training of information disseminators to be a good way to motivate other women to make the switch to green energy. Through this, a 'snowball' system can be initiated – which does, however, need some time to be effective on a large scale.

Project findings have led to several key recommendations that promote the purchase of green power and address obstacles and constraints experienced and analysed during the project. It should be highlighted that most findings of the project can also apply for the promotion of other consumption choices related to energy. In particular, the intervention involving gender-sensitive training, and personal communication and dissemination has proved to be effective and can be used, for instance, for the promotion of energy saving behaviour.

Efforts to promote green power among women should put an emphasis on issues that correspond to the most common motives of women, namely switching to green electricity as a contribution to phasing out nuclear power, protecting the environment, and influencing social developments in more general terms. How the green energy provider is using additional revenues to invest in renewable energies and which efforts are made to transform the energy system, including low-carbon development and phase-out of coal are vital topics of promotional activities. Successful promotion needs proper communication channels and trustworthy green power labels. Moreover, women need to be equipped with good arguments, as their male partners may obstruct their efforts to purchase green power, since the latter are often in charge of decision-making on issues related to energy.

The most promising target groups are the age groups between 30 and 50 who are most interested in green power, while other factors such as educational attainment play a minor role. The role of some other factors, such as rural or urban, family status and income are not completely clear. Yet, a low socio-economic status can impede the switch to green power, due to a lack of income, and also due to a lack of time or information.

Changing the electricity provider is a process that takes time, even if women have, in principle, already taken the decision to switch. This was confirmed by the survey and the intervention and should be taken into consideration for any promotional actions. Time constraints, in particular the double burden of employed women, a lack of impulses and the necessary efforts to find the "right" provider are among the underlying reasons for the "execution gap". Many women feel this to be a life decision, different from other purchases, ignoring that, in Germany, consumers can change their electricity provider repeatedly and at any time. To assist the participants of our training sessions to actually implement their decision, a booklet was prepared to serve as a "homework diary". It included FAQs, a short list of green power providers who are recommended by an alliance of German environmental NGOs promoting green power and hints to save electricity.

Extra costs are not the main barrier. In contrast to the assumptions at the beginning of the project, costs play a minor role for women. The higher price of green power is decisive only for those with a very low income. To address this barrier, options for other tariff models should be explored, for instance an exemption for welfare recipients from the extra "solar cent" which is charged by some green power providers. A more effective solution would be to introduce social tariffs in general, charging lower per kilowatt hour prices up to a certain consumption level, and in turn increasing specific prices for higher consumption.

Green power involves a number of peculiarities compared to consumption goods and products. The character of green power as a service, and the special conditions of electricity require specific interventions. First, there is a lack of noticeability, related to the fact that the utility of electricity is the same before and after the switch to green energy. In other words, a light bulb or washing machine powered by green electricity does not do its job better or healthier. The lack of visibility of green power can also impede the multiplier effect of interventions, as you can't show it and can't show off with it, different from, for example, a solar installation placed on the roof which demonstrates a household's environmental consciousness. Thus, a shift to green energy is ultimately driven by political principles and sustainable outcomes. Labels on letter boxes, windows and doors may be some compensation for this, but they ultimately have only a limited effect on outside perceptions. Green power purchasers could be shown in television programs which might raise awareness and remind hesitant switchers. Furthermore, electricity is not perceived as "consumption" such as buying organic or conventional apples, but rather as a subject matter of a contract. Consequently, it is more difficult to address and motivate people, as there are no emotions involved, except from exceptional circumstances, for instance after a nuclear accident.

A major obstacle preventing women from switching to green power is the complexity of the electricity market and the unsatisfactory organisation of the green power market. In Germany, several green power labels applying different criteria coexist. Moreover, there is a multitude of "green tariffs" offered by conventional utilities. In Berlin, for instance, consumers are confronted with 136 different tariffs offered by a large number of suppliers. Sixty percent of these tariffs are marketed as green power, most of them with a green power label. Even for experts, this is confusing and hard to assess and compare the quality of the various offers. We therefore recommend to policy-makers to define clear criteria and requirements for electricity products that are to be marketed as green power.

To properly respond to the specific character of the electricity market, interventions based on the multiplier effect of person-to-person communication appear to be advisable. Women have a high confidence in persons they know which might help to overcome the confusing market situation. Moreover, friends can act as reminders when the decision to switch has not yet been implemented. However, the training sessions have turned out to be a laborious option, and the women who took part might disseminate the information among friends and colleagues, but very few of them were willing to organise training sessions themselves, as this requires not only a high degree of engagement, but also a sufficient amount of time, which women do not always have. On the basis of our experiences, we would therefore recommend training energy experts in both general communications and gender-sensitive marketing and having them run training sessions for information disseminators, instead of turning lay people into energy experts who can carry out the training sessions. Existing structures, for instance women's local initiatives, mother's groups, and day care centres, can be encouraged to host such events.

In any case, the training sessions were very valuable in terms of their multiplier effect, even though this did not play out in the way we envisaged. As a recommendation, we would encourage the work-like nature of the training sessions to be reduced through a choice of other forms and locations and through the use of other, simpler concepts. As an example, a 'green energy gathering' which deviated in form, location and time from conventional training sessions was very well received. Combining entertainment with information provision, concrete support and/or advice throughout the process of switching to green energy would also be successful and able to be implemented in smaller style events. Another important information channel undoubtedly is the Internet. Yet, web sites to compare tariffs should be more transparent and provide more background information. Moreover, as women tend to be very active users of social networks, these should also be taken into consideration as potential dissemination channels.

More information (mainly in German)

www.genanet.de
www.strom-wechsel-frauen.de

References

Barletta M (2003) Marketing to Women: how to understand, reach, and increase your share of the world's largest market. Dearborn, Chicago

Birzle-Harder B, Götz K (2001) Grüner Strom – eine sozialwissenschaftliche Marktanalyse. ISOE (Studientexte, Bd. 9), Frankfurt a.M

Borgstedt S, Reusswig F, Christ T (2010) Umweltbewusstsein in Deutschland 2010. Ergebnisse einer repräsentativen Bevölkerungsumfrage. BMU und UBA, Berlin/Dessau

Deutsches Institut für Erwachsenenbildung (1994) Frauen(forschung) in der Erwachsenenbildung. Zeitschrift für Weiterbildungsforschung

Energie & Management (2011) Auf Rekordkurs. 7. E&M Ökostromumfrage. powernews, July 13. Retrieved 26 Oct 2011 from http://webcluster.montelpowernews.com/News/Story.aspx?id=91074&lang=G&UMMCat=126&Pub=0
European Commission (2009) Europeans' attitudes towards climate change (Wave 69.2 – TNS opinion & social No. Special Eurobarometer 300), European Parliament, European Commission, Brussels, Belgium, p 67
Faulstich-Wieland H, Willems K, Feltz N, Freese U, Läzer KL (2008) Genus – geschlechtergerechter naturwissenschaftlicher Unterricht in der Sekundarstufe 1 (1., Aufl.). Klinkhardt
Forsa (2007) Gesellschaft für Sozialforschung und statistische Analysen mbH: Meinungen zum Klimaschutz, Berlin
Gieseke W (2001) Handbuch zur Frauenbildung. Leske + Budrich, Opladen
Gindl M, Hefler G, Hellmer S (2007) Grundlagen der Gendersensibilität in der Lehre: Leitfaden für gendersensible Didaktik. MA 57 – Frauenabteilung der Stadt Wien, Wien
immowelt (2010) Wohnen und Leben 2010. Repräsentative Studie zu Wohnen und Leben in Deutschland
Iyengar SS, Lepper MR (2000) When choice is demotivating: can one desire too much of a good thing? J Personal Soc Psychol 79(6):995–1006
Johnsson-Latham G (2006) Do women leave a smaller ecological footprint than men? Initial study of lifestyles, consumption patterns, sustainable development and gender. Swedish Ministry of Sustainable Development
Johnsson-Latham G (2007) A study on gender equality as a prerequisite for sustainable development (Report to the Environment Advisory Council), Sweden
Kuckartz U, Rädiker S, Rheingans-Heintze A (2006) Umweltbewusstsein in Deutschland 2006. Ergebnisse einer repräsentativen Bevölkerungsumfrage. BMU, Berlin
Kuckartz U, Rheingans-Heintze A (2004) Umweltbewusstsein in Deutschland 2004. BMU, Berlin
Marktforschung.de (2011) Umdenken durch Fukushima: Mehrheit der Deutschen hält Energiewende für möglich. Martkforschung, May 10. Retrieved from http://www.marktforschung.de/information/nachrichten/marktforschung/umdenken-durch-fukushima-mehrheit-der-deutschen-haelt-energiewende-fuer-moeglich/
Schultz I, Stiess I (2009) Gender aspects of sustainable consumption strategies and instruments. Eupopp Working Paper 1. Institute for Social-Ecological Research (ISOE), Frankfurt/Main, Germany
Sunderer G (2006) Was hält Verbraucher vom Wechsel zum Ökostrom ab? Eine theoretische und empirische Analyse. Schriftenreihe des Zentrums für europäische Studien, Trier
Turner J, Grieco M (2000) Gender and time poverty: the neglected social policy implications of gendered time, transport and travel. Time Soc 9(1):129–136
Umweltbundesamt (2004) Quantifizierung der Effekte der Ökologischen Steuerreform auf Umwelt, Beschäftigung und Innovation (Hintergrundpapier), Umweltbundesamt für Mensch und Natur, Berlin, Germany, p 24
Underhill P (2000) Why we buy: the science of shopping. Simon & Schuster, New York
Weller I (2008) Konsum im Wandel in Richtung Nachhaltigkeit? Forschungsstand und Perspektiven. In: Lange H (ed) Nachhaltigkeit als radikaler Wandel. Die Quadratur des Kreises? VS Verlag, Wiesbaden, pp 43–70
Weller I, Hayn D, Schultz I (2002) Geschlechterverhältnisse, nachhaltige Konsummuster und Umweltbelastungen. In: Balzer I, Wächter M (eds) Sozial-ökologische Forschung. Ergebnisse der Sondierungsprojekte aus dem BMBF-Förderschwerpunkt. Ökom, München, pp 431–452

Chapter 11
A Path to Implementation: Gender-Responsive Climate Change Strategies

Lorena Aguilar

Abstract Climate change is one of the most urgent priorities on the global agenda. The impacts are already being felt, with every country in the world grappling with the challenges of mitigating the causes and—especially for developing countries and vulnerable populations—adapting to its effects. Increasingly, the evidence reveals that the impacts of climate change are not gender-neutral. Women and men experience climate change differently and their capacity to cope with it varies.

The 2007 UNDP Human Development Report cautions that gender inequalities intersect with climate risks and vulnerabilities, concluding that climate change is likely to amplify and exacerbate existing patterns of gender disadvantage. Thus, women's historic disadvantages, their limited access and control over decision-making, environmental and economic resources, and their restricted rights, make them more vulnerable to climate change.

Keywords Gender • Climate change • Developing countries • Environment • Economics

Recognizing Women as Part of the Solution

Climate change is one of the most urgent priorities on the global agenda. The impacts are already being felt, with every country in the world grappling with the challenges of mitigating the causes and—especially for developing countries and vulnerable populations—adapting to its effects. Increasingly, the evidence reveals that the impacts of climate change are not gender-neutral. Women and men experience climate change differently and their capacity to cope with it varies.

L. Aguilar, M.Sc. (✉)
IUCN (International Union for the Conservation of Nature), Gland, Switzerland
e-mail: lorena.aguilar@iucn.org

M. Alston and K. Whittenbury (eds.), *Research, Action and Policy: Addressing the Gendered Impacts of Climate Change*, DOI 10.1007/978-94-007-5518-5_11,

The 2007 UNDP Human Development Report cautions that gender inequalities intersect with climate risks and vulnerabilities, concluding that climate change is likely to amplify and exacerbate existing patterns of gender disadvantage. Thus, women's historic disadvantages, their limited access and control over decision-making, environmental and economic resources, and their restricted rights, make them more vulnerable to climate change.

However, this disproportionate burden of climate change on women can be countered by their empowerment and recognition. Women have important primary roles as managers of forests, land, water, and other natural resources in many communities. This position makes them **powerful agents of change** in formulating responses to climate change. Women are part of the solution.

Women act as leaders in identifying ways in which to cope with and adapt to climate. They are fundamental to mitigating climate change, largely due to their critical role in optimizing energy efficiency, their receptiveness to greener sources of energy, as well as their power to change and influence consumption patterns.

Thus, national and regional climate change strategies will have a much greater potential to ensure efficiency, effectiveness and quality of implementation if they are developed to include a gender component, paying particular attention to the role and contributions of women.

We've Come a Long Way

Despite all this, gender was absent from decisions by its Conference of the Parties (COPs) or its Subsidiary Bodies for the entire first two decades of the United Nations Framework Convention on Climate Change (UNFCCC). In addition, very few National Adaptation Programmes of Action to Climate Change (NAPAs) and National Communications submitted by Parties to the UNFCCC addressed gender considerations. Recognizing this glaring imbalance between deficiency and potential, IUCN's Gender Office has been working with its partners to ensure that gender considerations are acknowledged in climate change discussions, policies and strategies at the international, regional and national level.

Focusing on knowledge development, capacity building and advocacy, IUCN's work on gender and climate change has already yielded considerable impact. Some successful outcomes include:

- Co-founding the Global Gender and Climate Alliance (GGCA), an alliance of 60 UN agencies and civil society institutions working together on gender and climate issues.
- Publishing of the first Gender and Climate Change Training Manual in 2008 (along with other key knowledge products) now used around the world to include a gender perspective in the analysis and implementation of national and regional adaptation and mitigation measures.
- Training nearly 500 experts from more than 100 countries on the linkages between gender and climate change. Trainings have created a cadre of experts

that are available and fully prepared to engage in the climate change community, and many trainees have used the knowledge gained to spearhead their own initiatives in their home countries, thereby multiplying the impact of GGCA efforts.

- Enabling women as part of the GGCA efforts, and under the guidance of the Women Environment and Development Organization (WEDO), from the global South to participate as members of their nations' UNFCCC negotiating team. Sponsorship by the Women Delegates Fund increased the number of women participants at major UNFCCC meetings—to date, 85 trips have been funded for 28 delegates to attend more than ten global meetings. Some of the sponsored women delegates represented the only woman on their national delegations.
- Providing orientation sessions on the integration of gender and climate change for government delegates from nearly 110 countries. These sessions became a driving force behind government representatives demanding the development of gender-sensitive policies and initiatives at the national level.

At the global level, governments have now widely agreed that gender equality is a key aspect to achieving climate change goals and other development targets. Governments are now increasingly paying special attention to gender issues, with more than 40 official references made to gender in UNFCCC negotiations since 2008. The most recent set of UNFCCC decisions towards a global deal, known as the Cancun Agreements, contains no less than nine strong gender references, and the Durban Outcomes comprises seventeen references related to gender.

Development of Gender-Sensitive Strategies[1]

This paradigm shift is in no small measure the result of the continued efforts of IUCN and its partners. The understanding of the linkages between gender and climate change has resonated with governments which now want to mainstream gender in their national climate policies and plans. IUCN is working with these governments, many of which are IUCN members, to build their capacity to develop gender-responsive climate change strategies and programmes.

Methodological Process

Figure 11.1 illustrates the process and key elements of developing a national or regional gender-sensitive climate change strategy by IUCN and its national partners.

---

[1] These gender-responsive climate change strategies have been produced thanks to the financial and visionary support provided by the government of Finland.

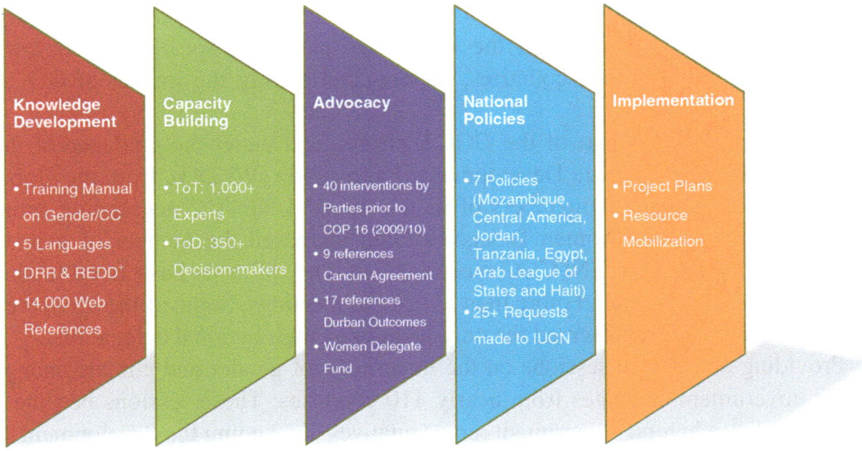

Fig. 11.1 Gender and climate change process (Notes: *DRR* disaster risk reduction, *ToT* training of trainers, *ToD* training of delegates)

Although firmly structured and detailed, the method remains importantly responsive to the differing socio-political and environmental contexts within which it is applied. As the figure shows, implementation of plans and strategies flow from national policies, which are grown out of global, national and local advocacy. The entire process rests upon a fundamental, initial stage of comprehensive knowledge and capacity building on gender and climate change issues.

Characteristics of the Strategies

Although tailor-made to the needs and circumstances of each country, each IUCN-developed national/regional strategy has a set of generic, defining characteristics. They:

- Are demand-driven.
- Cover a wide range of topics, as agreed by stakeholders.
- Provide solutions to both rural and urban communities.
- Require resultant projects and initiatives to have concrete indicators and to be built from the bottom-up.
- Provide technical solutions at all levels, from the household to national level.
- Are flexible in that when further policy modification is required, appropriate revision is possible.
- Contain institutional imperatives.
- Build the capacity and coordination in and amongst institutions related to climate change.

In being driven and led by countries themselves, the national strategies are built on the capacity of local stakeholders and ensure local ownership; are

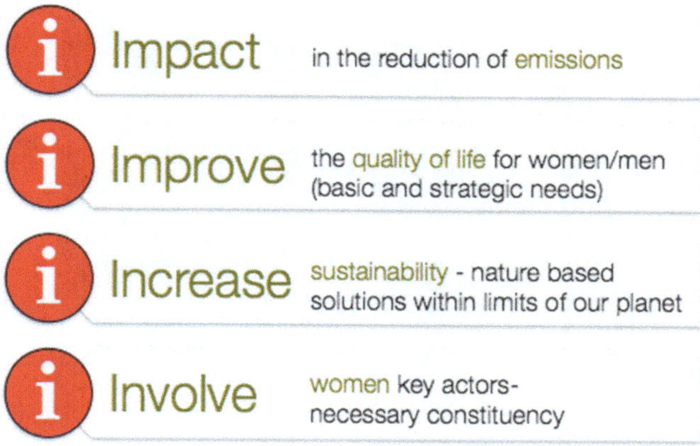

Fig. 11.2 Principles behind the strategies dealing with mitigation

entrenched in local context; respond to local needs; suit internal institutional arrangements; and undergo national legalization processes—securing ultimate success in implementation.

The strategies are based on firm principles which focus their development on obtaining results—as shown in Fig. 11.2—aiming to ensure that national and regional climate change processes mainstream gender considerations in order to guarantee that women and men can have access to, participate in and benefit equally from climate change initiatives.

The process though which national strategies are developed necessitates careful analysis of the situation of gender and climate change, conducted by the local stakeholders themselves, and operational planning for the future. As shown in Fig. 11.3, this analysis and planning is comprehensive and cross-sectoral, examining key issues such as consumption, energy, forests, waste and transport through the lenses of behavioural change, technology, capacity building and awareness-raising in relation to mitigation. The strategies also cover adaptation topics such as water, disasters, agriculture and nutrition, land degradation and amongst others.

With this approach, IUCN's engagement efforts have led to the development and support of regional and national plans of action for climate change that crucially include gender dimensions.

Recent Successes in Establishing National and Regional Strategies: Some Examples

Mozambique

In January 2010, the IUCN Gender Office in Alliance with UNIFEM developed the world's first gender-sensitive climate strategy in Mozambique. The National Climate Change and Gender Strategy incorporates gender mainstreaming and

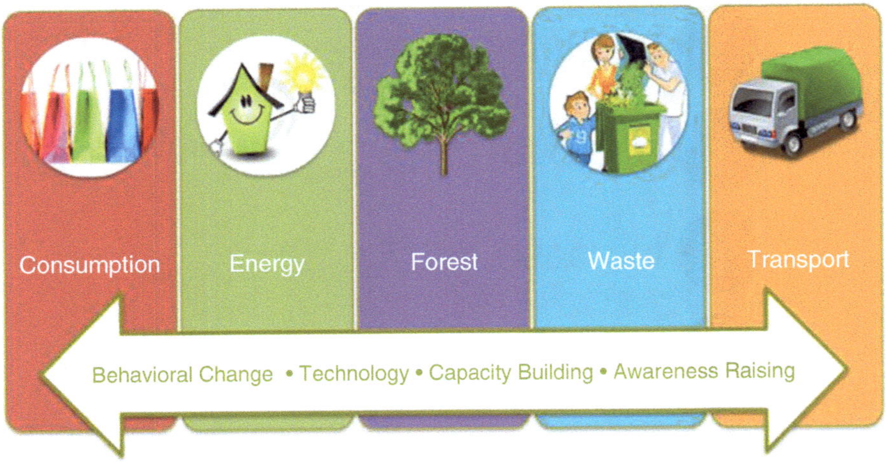

Fig. 11.3 Mitigation topics covered by the national strategies

includes gender indicators for all areas of climate change activity undertaken by the national government. The process of developing the strategy in Mozambique raised women's voices and concerns into an arena of high national priority and established the unprecedented expectation that all climate change related projects and activities conducted in the county by donors, NGOs, and external partners will incorporate gender considerations.

The strategy has undergone a national legalization process: it was submitted for comment at provincial level, followed by its presentation to the National Council for Sustainable Development, where it was assessed positively. It was later adopted and endorsed by the Council of Ministers. The strategy for Mozambique, with its respective Plan of Action sets out priorities in harmony with other major national environmental actions such as the PECODA, NAPA and—importantly for forests—the Action Plans for Combating Erosion and Uncontrolled Fires, and the Presidential Initiatives "One Student One Tree Each Year" and "One Leader One Community Forest." In the roll-out plan for implementing the strategy, the Ministry for the Coordination of Environmental Affairs has the objective of reaching around 1.5 million communities in a 5-year period (2010–2014) and to cover the entire country by 2025.

Central America

During the summer of 2010, IUCN in collaboration with the Regional Unit for Technical Assistance (RUTA) and CoopeSoliDar finalized the first gender-sensitive climate change strategy for the Central American region. Recognized as one of the most biodiversity-rich regions in the world, Central America is also moving towards a political integration system—Sistema de la Integración Centroamericana

(SICA)—whereby all of the Ministers of Environment in the region fall under one Central American Commission on Environment and Development (CCAD), a coordinating body working on developing a regional climate change strategy. Under SICA, civil society is recognized and women are acknowledged as a major group, and they can therefore participate and engage fully in advocacy.

The strategy provides regional guidance and was upheld at a Meeting with Network of Women Ministers (COMCA), approved and legalized by the Council of Ministers of the CCAD. It created such political pressure that it led to the incorporation of gender as a mandate in the Heads of State declaration in relation to the climate change strategy for the region (July 2010).

Jordan

In the autumn of 2010, IUCN supported and developed the "Programme for Mainstreaming Gender in Climate Change Efforts in Jordan", a key document used by the Government of Jordan in the development of the country's Third National Communication to the UNFCCC. The subsequent Plan of Mainstreaming Action on Gender and Climate Change aims to ensure that national climate change efforts mainstream gender considerations so that women and men can have access to, participate in, contribute to and hence optimally benefit from climate change initiatives, programs, policies and funds.

Egypt

In June of 2011, IUCN supported the development of the National Strategy for Mainstreaming Gender in Climate Change in Egypt. The framework for integrating a gender perspective in climate change efforts in Egypt covers the period 2011–2016 and again will be an integral part of the Third National Communication Report for the country. It establishes objectives, outlining substantive activities with achievable indicators within the ambit of eight priority sectors. The objective of this Strategy is to mainstream gender considerations into national climate change initiatives and policies, so that both men and women have equal opportunities—as in Jordan's strategy—ultimately contributing to national economic, environmental and social sustainability.

United Republic of Tanzania

Most recently, and at the time of writing, IUCN has been supporting the government of Tanzania to develop a national strategy on gender and climate change. The multi-stakeholder process of developing the gender strategy is aligned with the development of the National Climate Change Strategy and Action Plan.

Elsewhere . . .

In addition to these examples, IUCN is in the process of finalizing similar strategies for Costa Rica, Haiti, Liberia, Nepal and Panama. IUCN is trying to respond to the demand of more than 25 governments from all regions of the world that have requested support to develop national plans of actions on gender and climate change.

Leila Nafa'a Hamarneh, Projects Director, Arab Women Organization, Jordan

I have been involved in implementing programmes and projects related to women's issues and women's rights at the Arab Women Organization, a leading civil society organization in Jordan. A GGCA training workshop addressing gender and climate change held by IUCN in Amman in 2009, and the GGCA 'Training Manual on Gender and Climate Change' awoke me to the importance of linking gender and climate change. The workshop made me realize that women should address broad issues pertaining to development, sustainable livelihoods and poverty eradication in addition to calling for anti-discrimination reforms and combating violence against women. The Training Manual wove the two topics into an elegant fabric and illustrated how international standards can serve to maintain gender equality and sustainable livelihoods.

Prior to the Copenhagen Conference, the GGCA, IUCN, the Arab Women Organization and civil society led an advocacy campaign to introduce gender into Jordan's National Third Communication to UNFCCC. The campaign's impact was so effective that Jordan was among the few to consider gender within its work plans. After the Copenhagen Conference, the Arab Women Organization was invited to get involved in the preparation of an Action Plan for Gender and Climate Change.

The training on gender and climate change and the experiences I gained while being involved in preparing a gender and climate change strategy have given civil society and the women's movement a rare opportunity to learn and leave footprints in Jordanian policies and strategies. The Arab Women Organization received reports from two women activists who were trained with the GGCA Training Manual, saying that they insisted on introducing the gender and climate change perspective while discussing Risk Management at a Jordanian Red Crescent Conference and while discussing sustainable agriculture in the Ministry of Agriculture.

Stimulating Cooperation and Opportunity for Other Sectors

The national strategies create coherence and bridge the divides between different government departments and other stakeholders dealing with women and climate change. Through the multi-stakeholder engagement process, the strategies increase public awareness of climate change and gender equality issues, to stimulate interest, participation, and consultations with different stakeholders in developing climate change policy, and to increase capacity of those national actors to continue their involvement.

Chapter 12
The Gender Gap in Environmental Attitudes: A System Justification Perspective

Rachel E. Goldsmith, Irina Feygina, and John T. Jost

Abstract System justification refers to a psychological tendency to maintain certainty, security, and solidarity through motivated perceptions of the status quo and the extant socioeconomic system as beneficial, fair, stable, and legitimate, especially in response to dependency and threat. System justification impedes efforts to address societal challenges, and in particular gives rise to denial, resistance, and inaction in the face of climate change and environmental problems. Women chronically engage in less system justification than men, and this difference partially explains women's greater willingness to acknowledge ecological problems and risks and to engage in actions that are beneficial for the environment. We demonstrate that reframing environmental messages as consistent with upholding the established way of life and the well-being of our society gives rise to increased support for environmental efforts on the part of those who are especially motivated to justify the system and can therefore help to narrow the ideological gap in environmental attitudes and behaviors.

Keywords Gender • Environmental attitudes • Environmental behaviours • Climate change • System justification

A widespread and consistent finding reveals that women, as compared to men, demonstrate higher levels of awareness and response to environmental problems (e.g., McCright 2010; World Bank 2009). Gender differences have been reported in pro-environmental attitudes, concern for the environment, knowledge levels

R.E. Goldsmith (✉)
Rush University Medical Center, Chicago, IL, USA
e-mail: RachelEvelynGoldsmith@gmail.com

I. Feygina • J.T. Jost
New York University, New York, NY, USA

M. Alston and K. Whittenbury (eds.), *Research, Action and Policy: Addressing the Gendered Impacts of Climate Change*, DOI 10.1007/978-94-007-5518-5_12,

regarding climate change, and willingness to take action to improve environmental problems (Bord and O'Connor 1997; Semenza et al. 2011; Zelezny et al. 2000). How can we understand these and related differences?

Gender and Psychological Responses to Climate Change

It is impossible to understand current responses to the environmental crisis, as well as gender differences in these responses, without taking into consideration the impact of social and psychological processes. Research reveals that cognitive, affective, interpersonal, and societal factors influence environmental knowledge, concern, and action (Swim et al. 2011). Barriers to acknowledging and addressing environmental problems include challenges in procuring accurate information; assessing the likelihood, severity, and causes of environmental threats; feelings of separation or alienation from natural ecosystems; prioritizing individual gain over collective well-being; discomfort with uncertainty; attachment to habits; perceptions of problems arising from climate change as temporally and geographically distant; and the difficulty of collaborating with others to tackle ecological problems effectively (e.g., Antal and Hukkinen 2010; Gardner and Stern 2002; Gifford 2011; Swim et al. 2009; Takacs-Santa 2007). Explanations aimed at elucidating gender differences in support for the environment have posited women's stronger "ethic of care", more prevalent altruistic tendencies, greater capacity for perspective-taking, and stronger empathic tendencies (Arnocky and Stroink 2011; Dietz et al. 2002; Zelezny et al. 2000).

In addition to these factors, our research addresses the role of motivational processes in the formation and change of attitudes and responses concerning environmental problems. Specifically, we examine how the motivation to justify the socioeconomic system and status quo, especially but not exclusively in response to threat (e.g., Jost et al. 2004; Jost and Hunyady 2005), influences responses to ecological challenges (Feygina et al. 2010a, b). Moreover, we focus on how the widespread impetus to rationalize "the way things are" affects the relationship between gender and environmentalism.

Gender Differences in Environmental Knowledge, Concern, and Action

We begin by addressing the scope of gender differences in knowledge regarding global climate change, levels of concern for the environment, and extent of engagement in pro-environmental actions, and then consider prior explanations that have been offered to account for these gender differences. Relationships between gender and environmental beliefs and actions have been investigated in many countries, settings, and age groups. Overall, women display greater pro-environmental attitudes and more concern and willingness to take action to help the environment, compared to men (Carrier 2007; Davidson and Freudenburg 1996; Dietz et al. 2002; Stern et al. 1993; Sundblad et al. 2007; Wang 1999; Zelezny et al. 2000). Differences are also observed with respect to knowledge regarding the science of climate change as well as self-perceptions about knowledge of environmental issues. For instance, U.S. surveys conducted by the Pew Research Center reveal that women are more likely than men to believe that there is "solid evidence" for global warming (Egan and Mullin 2012). Similar findings emerge from Gallup Polls surveys collected over the last decade, which assess opinions about whether pollution from human activities constitutes the primary cause of global warming, whether most scientists believe global warming is occurring, and whether the effects of global warming have already begun to happen. Statistically adjusting for other factors, including political ideology, party affiliation, educational attainment, age, race, and annual income, women were more knowledgeable about climate change than were men, and this gender difference remained consistent over time. However, adjusting for climate change knowledge and related variables, women were more likely than men to underestimate their scientific knowledge of climate change (McCright 2010). Finally, in an international World Bank (2009) study of public attitudes regarding climate change, women in the U.S., France, Russia, Japan, and Bangladesh were more likely than men to believe that climate change has already begun to harm people. Men were more likely than women to hold this belief in only one country, Iran.

Women also express higher levels of concern and perceptions of risk regarding environmental issues, compared to men (Arnocky and Stroink 2011; Bord and O'Connor 1997; Davidson and Freudenburg 1996; Dunlap and McCright 2008; Riechard and Peterson 1998; Tranter 2011; Zelezny et al. 2000). These gender differences remained statistically significant even after adjusting for environmental knowledge, religiosity, and socialization (Zelezny et al. 2000). Women also report more concern about the effects of climate change than do men (Brody et al. 2008; Hamilton 2008; Leiserowitz 2006; Malka et al. 2009; McCright 2010; O'Connor et al. 1999), a difference that persists even after adjusting for employment status, parenthood, and being a full-time homemaker (McCright 2010).

Studies of environmental *action* also reveal that women are more likely than men to actively participate in pro-environmental behaviors and to express willingness to contribute financially to protecting the environment (Tranter 2011; Wehrmeyer and McNeil 2000). For example, women are more likely than men to report reducing energy consumption in response to information about climate change (Semenza et al. 2011). Women from 20 different nations were found to participate in significantly more environmentally-oriented behaviors than men in the private sphere (e.g., recycling, transportation choices) but not in the public sphere (e.g., volunteering, attending meetings or protests) (see Hunter et al. 2004). In research involving undergraduate samples from the U.S., Europe, and Latin America, women reported greater pro-environmental action in 11 of 14 countries (Zelezny et al. 2000), with gender differences more pronounced for behavioral than attitudinal measures. Finally, undergraduate women display greater willingness than men to engage in ecological cooperation within the context of a commons dilemma (Arnocky and Stroink 2011).

Explaining Gender Differences in Environmentalism

Several explanations have been proposed for gender differences in environmentalism.

Zelezny et al. (2000) argued that women, compared to men, are more likely to take the perspective of others, and to hold a stronger "ethic of care." Across cultures, women tend to be socialized to be more expressive, nurturing, cooperative, and caregiving than men, whereas men are socialized to be more competitive and independent than women. Being ecocentric – concerned with environmental needs – suggests an "other" orientation, which, according to gender socialization theory, is more characteristic of female than male socialization. In line with this argument, women from 14 different countries not only reported greater environmental concern, but also expressed a stronger sense of social responsibility and affinity for taking others' perspectives, as compared to men in those same countries.

Gender differences in altruism, empathy, and cooperation may also help to illuminate the nature of the relationship between gender and environmentalism. Female undergraduates are more likely than males to be concerned about the negative impact of environmental problems on personal well-being, social welfare, and the sustainability of the planet, which contributes to a greater willingnes to take political action and pay higher taxes to support the environment (Stern et al. 1993). Moreover, the greater value women ascribe to altruism seems to underlie their greater endorsement of pro-environmental attitudes (Dietz et al. 2002). Finally, women's capacity for empathy contributes to increased willingness to act cooperatively in an environmental commons dilemma (Arnocky and Stroink 2011).

Impact of System Justification Tendencies on Environmentalism

While the findings discussed above highlight important differences between women and men that contribute to gender differences in environmental attitudes and behaviors, in our research we explore an additional contributor: the motivation to justify the existing social, economic, and political system and status quo. *System justification* refers to the psychological tendency to maintain the perception of social stability, certainty, security, and belongingness through motivated perceptions of the status quo and the extant socioeconomic system as fair, legitimate, and desirable (Jost et al. 2004, 2008a; Jost and Hunyady 2005). In the short term, system justification assuages anxiety, fear, and uncertainty that arise when the status quo is threatened (e.g., Jost and Hunyady 2002). However, in the long term, system justification can inhibit the pursuit of positive social change and collective action aimed at correcting injustices and system-level problems (Wakslak et al. 2007). It may produce especially negative consequences for members of disadvantaged groups, including the perpetuation of inequality, stigma, and other structural barriers to well-being (e.g., Jost and Thompson 2000; O'Brien and Major 2005; Rankin et al. 2009).

Applying system justification theory to the scientific understanding of responses to climate change suggests multiple layers of interpretation. The model highlights defensive psychological processes that arise when individuals are confronted with the threat of climate change, and it speaks to empirical connections involving gender, political ideology, and attitudes concerning the environment. Current environmental problems are bound to prevalent socioeconomic institutions and practices (Axelrod and Suedfeld 1995; Shrivastava 1995; White 1967). These include industrial practices and technologies that drain the Earth's resources and create pollution that endangers ecosystems and human health, as well as political indifference to environmental harm, and cultural assumptions about dominating and exploiting nature for the sake of progress and growth. Thus, facing up to environmental problems and their anthropogenic causes necessitates finding fault with and questioning many aspects of the societal status quo, which conflicts with psychological needs to perceive the social system as legitimate and stable. We propose that, as a result, information about climate change and proposals to take action are likely to instigate defensive responses from individuals and groups who are motivated to justify the system.

The conflict between negative information regarding climate change and the motivation to preserve current social, economic, and political arrangements creates psychological "dissonance" which people are motivated to reduce (Festinger 1957) by minimizing or denying one of the conflicting cognitions, while generating increased support for the chosen belief. While threats from outside (e.g., international terrorism) often elicit aggressive responses, the environmental challenge is internal to the system. As a result, we propose that defensive mechanisms will often result in minimization, distortion, and denial of the problem and of human responsibility for it.

Several prior studies support these claims, and demonstrate that psychological defense mechanisms contribute to apathetic responses to the problem of climate change (e.g., Hollander 2009; Moran 2011; Norgaard 2011). Denial (the rejection of reality) is used to resolve dissonance between concerns about climate change and the desire to maintain continuity in one's lifestyle (Stoll-Kleemann et al. 2001). In addition, climate change frequently gives rise to sadness, disgust, guilt, anxiety, helplessness, and depression. Denial is used as a psychological defense to buffer the experience of these negative feelings (Doherty and Clayton 2011). Similar defense mechanisms guard against distress from environmental degradation, including "rational distancing" (i.e., the cutting off of emotional responses) and diffusion of responsibility (rejecting personal responsibility and blaming others for environmental problems; see Kollmuss and Agyeman 2002). Such defense mechanisms are likely to interfere with taking effective action against the deleterious consequences of climate change (Hollander 2009). Finally, environmental denial appears to increase with investment in current socioeconomic practices. In one of many examples, individuals who profess stronger levels of support for a capitalist economy are less likely to believe that climate change is occurring, that it is caused by human activity, and that its consequences are negative (Heath and Gifford 2006).

Empirical Evidence

To provide even more direct evidence for the proposed relationship between system justification and denial of environmental realities we conducted a series of survey and experimental studies (see Feygina et al. 2010b). First, we assessed the extent to which our participants were motivated to justify the societal status quo by asking whether they agreed or disagreed with statements such as "Most policies serve the greater good," "Society is set up so that people usually get what they deserve," "In general, the American political system operates as it should," and "American society needs to be radically restructured" (Kay and Jost 2003). Participants reported their environmental attitudes using the *New Environmental Paradigm* scale (NEP; Dunlap et al. 2000), which assesses: denial of the possibility of an ecological crisis (e.g., "If things continue on their present course, we will soon experience a major environmental catastrophe"); denial of limits to growth (e.g., "The earth has plenty of natural resources if we just learn how to develop them"); denial of the need to abide by the constraints of nature (e.g., "Humans will eventually learn enough about how nature works to be able to control it"); and denial of the danger of disrupting balance in nature (e.g., "The balance of nature is strong enough to cope with the impacts of modern industrial nations"). We found that, as hypothesized, the more participants were invested in the system and motivated to uphold it through justification, the more they engaged in all four types of denial of environmental realities.

System Justification Contributes to Gender Differences in Environmentalism

These findings also have direct implications for explaining gender differences in environmentalism. The motivation to justify the system is related to an individual's status within society. Occupying a privileged position within a hierarchical social order yields many benefits. Women, minorities, and the poor are often in a situation of disadvantage, while groups who control material resources exploit those lacking resources (Kendall et al. 2006). The advantaged are especially motivated to maintain and justify current systems and are typically more likely to engage in system justification (Jackman 1994; Jost and Thompson 2000; Rankin et al. 2009; Sidanius and Pratto 1999). For instance, males typically score significantly higher than females on measures of system justification (Jost and Kay 2005). On the basis of this finding, we hypothesized that higher system justification among men would account for their greater denial of environmental realities, in comparison with the less privileged social status of women.

This idea resonates with prior theorizing, such as feminist frameworks that have offered similar explanations for the association between gender and environmental attitudes. Because of longstanding experiences of oppression, women more readily identify with the exploitation of natural resources by humans and have a stake in ending this exploitation (Shiva 1989). Both the feminist and environmental movements seek to establish egalitarian, non-hierarchical systems. In addition, feminist consciousness – which involves an awareness of social inequalities based on gender and a commitment to overcoming such disparities (Conover and Sapiro 1993) – is related to stronger support for the environment, among both women and men (Gupte 2002). Conversely, endorsement of right-wing authoritarianism – a preference for hierarchy and dominant leadership, as well as support for discrimination, hostility, and prejudice against out-groups (e.g., Altemeyer 1998; Whitley 1999) – is associated with less environmental support among men and women alike (Wang 1999). A system justification framework helps to integrate these findings, insofar as feminist consciousness and support for the environment both reflect a willingness to acknowledge and respond to the drawbacks of the status quo and the need to care for those who are vulnerable, whereas right-wing authoritarianism bolsters entrenched hierarchical systems that are consistent with the exploitation of the Earth and its resources.

Results from our studies (Feygina et al. 2010b) demonstrate that, as hypothesized, women engage in less system justification than men, and men are indeed more likely than women to deny environmental problems. Using mediational analyses (Baron and Kenny 1986), we determined that the gender difference in system justification statistically explains the gender gap in environmental attitudes, at least in part. In other words, men's greater denial of environmental realities is partially accounted for by their more enthusiastic endorsement of system justifying beliefs.

Connections Among Gender, Social Status, Ideology, and Environmentalism

Research also demonstrates that endorsement of system justification differs across the ideological spectrum. Political conservatives, compared to liberals, report stronger tendencies to justify the system (Jost et al. 2003, 2008b), and therefore engage in greater denial of environmental problems (Feygina et al. 2010b). A multitude of investigations have documented stronger tendencies to deny and minimize problems associated with anthropogenic climate change among political conservatives vs. liberals and moderates in the U.S. (Begley 2007; Saad 2007), Australia (Tranter 2011) and in two dozen other countries (Tjernström and Tietenberg 2008).

It is important to consider how these ideological factors interact with and magnify the effect of social status on gender differences in environmentalism (e.g., McCright 2010). McCright and Dunlap (2011) analyzed public opinion data from ten Gallup surveys, conducted between 2001 and 2010, which assessed five types of responses to climate change related to denial. These included beliefs about if and when the effects of climate change would occur; whether climate change is primarily caused by human activities; whether scientists have reached a consensus on climate change; the extent to which global warming is exaggerated in the news; and levels of worry about climate change. Results demonstrated that conservative White males were far more likely than other groups to perceive minimal risks in all five areas. For instance, 29.6% of conservative White males expressed the belief that the effects of global climate change will never occur, whereas only 7.4% of other respondents endorsed this view. A strong majority of conservative White males (65.1%) believed that the media has exaggerated the seriousness of global warming, whereas only 29.9% of other respondents held this view. Similar outcomes were observed for beliefs regarding scientific consensus, the role of human activities, and levels of worry. Interestingly, White males were more likely than other groups to report that they understood climate change very well, and self-reported understanding was positively correlated with the denial of climate change. McCright and Dunlap (2011) argue that these results reflect tendencies to engage in system justification (Feygina et al. 2010a, b), insofar as conservative White males are more likely to hold positions of political and economic power, and are therefore more motivated to defend the societal status quo. Exaggerated confidence in perceived understanding, and its association with denial of climate change (McCright and Dunlap 2011), may reflect fairly successful efforts to reduce cognitive dissonance by minimizing and rejecting undesirable information. This tendency to maintain consistent beliefs through motivated ignorance of aspects of reality presents a formidable barrier to addressing climate change (e.g., Antal and Hukkinen 2010; Kollmuss and Agyeman 2002; Shepherd and Kay 2012).

Similar correlations between gender and support for the environment are observed at the level of mass politics. Increased representation of women in national parliaments was associated with environmental treaty ratification in a comparison involving 130 countries that comprise of 92% of the world's population (Norgaard and York 2005). This suggests that the egalitarianism of societies and the inclusion of women in government are important factors in determining political support for pro-environmental policies.

Harnessing System Justification Motivation

Given the formidable barrier that system justification poses for acknowledging and responding to environmental problems, we sought to determine whether defensive psychological responses could be redirected toward building support for the environment and for positive structural change. In an experimental setting, we reframed environmentalism to emphasize the ways in which it is consistent with the well-being and perpetuation of the system: ("Being pro-environmental allows us to protect and preserve the American way of life. It is patriotic to conserve the country's natural resources"). We found that exposure to this message attenuated, and even reversed, the negative impact of system justification on environmental attitudes and behaviors. Among participants who exhibited strong tendencies to justify the system, those who were exposed to the system-congruent message reported greater intentions to help the environment through private and public actions, and were more likely to sign petitions intended to benefit the environment. In other words, we harnessed the motivation to uphold and protect the status quo and directed it toward helping, rather than harming, the environment. These results suggest that reframing environmental messages to address the needs of those who are strongly motivated to justify the system may be a powerful tool for overcoming resistance. We recommend that communications draw attention to the fundamental ways in which pro-environmental behavior is consistent with maintaining the well-being of society (e.g., the "American way of life"), in order to increase participation among individuals with system-justifying proclivities.

Our recommendation aligns well with others that have been proposed in light of evidence that purely science-focused approaches to communicating the importance of climate change to public audiences have been ineffective (Antal and Hukkinen 2010). Rather, environmental communication should target social, cognitive, and motivational dynamics through messages that link individual concerns about safety to broader issues of cultural survival, and pair depictions of environmental problems with convincing methods of addressing them (Antal and Hukkinen 2010). Strategies such as these are especially important in the U.S. and in Western Europe, where individuals perceive environmental issues to be related to ambitious, abstract goals such as social justice and world peace. They may be less necessary in countries like Japan, where environmental priorities appear connected to longstanding, traditional, proximal, and personal constructs, such as family security and honoring one's parents (Aoyagi-Usui et al. 2003). Messages regarding climate change can be made more effective by linking policy goals to local environmental contexts and personal concerns. Given that, as we have seen, women are typically more likely to concern themselves with the personal and collective impact of environmental problems (Stern et al. 1993), messages that stimulate men's interest and motivation in addressing environmental challenges would be especially valuable.

Conclusions

The research described in this chapter addresses the recent call by social scientists to investigate psychological barriers to the acknowledgment of climate change and commitment to pro-environmental action (e.g., Antal and Hukkinen 2010; Clayton and Brook 2005; Feinberg and Willer 2011; Swim et al. 2011). It also provides insight into the role of motivated system justification processes with respect to the widely observed gender gap in environmental attitudes. We find that the more individuals are invested in the status quo, and the more motivated they are to justify and uphold extant systems, the less willing they are to admit and confront environmental problems. To the extent that men traditionally occupy higher status in society than women, they benefit more from established social, economic, and political institutions and arrangements, exhibit stronger motives to justify the system, and are less willing to admit and tackle environmental challenges. We have shown that psychological barriers to confronting ecological crises can be countered by reframing environmental messages as consistent with defending and preserving the status quo. Our collective efficacy in addressing the environmental challenge may hinge upon the willingness to acknowledge our detrimental attachment to the status quo, and our ability to foster a psychologically sustainable approach to inspiring change.

References

Altemeyer B (1998) The other "authoritarian personality". In: Zanna MP (ed) Advances in experimental social psychology, vol 30. Academic, New York, pp 47–92

Antal M, Hukkinen JI (2010) The art of the cognitive war to save the planet. Ecol Econ 69:937–943

Aoyagi-Usui M, Vinken H, Kuribayashi A (2003) Pro-environmental attitudes and behaviors: an international comparison. Hum Ecol Rev 10:23–31

Arnocky S, Stroink ML (2011) Gender differences in environmental concern and cooperation: the mediating role of emotional empathy. Curr Res Soc Psychol 16:1–14

Axelrod LJ, Suedfeld P (1995) Technology, capitalism, and Christianity: are they really the three horsemen of the eco-collapse? J Environ Psychol 15:183–195

Baron RM, Kenny DA (1986) The moderator-mediator variable distinction in social psychological research: conceptual, strategic, and statistical considerations. J Pers Soc Psychol 51:1173–1182

Begley S (2007) The truth about denial. Newsweek. Retrieved on 14 Aug 2007 from http://www.msnbc.msn.com/id/20122975/site/newsweek/-COVER

Bord RJ, O'Connor RE (1997) The gender gap in environmental attitudes. Soc Sci Q 78:830–840

Brody SD, Zahran S, Vedlitz A, Grover H (2008) Examining the relationship between physical vulnerability and public perceptions of global climate change in the United States. Environ Behav 41:72–95

Carrier SJ (2007) Gender differences in attitudes toward environmental science. Sch Sci Math 107:271–278

Clayton S, Brook A (2005) Can psychology help save the world? A model for conservation psychology. Anal Soc Issues Public Policy 5:1–15

Conover PJ, Sapiro V (1993) Gender, feminist consciousness, and war. Am J Political Sci 37:1079–1099

Davidson DJ, Freudenburg WR (1996) Gender and environmental risk concerns. Environ Behav 28:302–339

Dietz T, Kalof L, Stern PC (2002) Gender, values, and environmentalism. Soc Sci Q 83:351–364

Doherty TJ, Clayton S (2011) The psychological impacts of global climate change. Am Psychol 66:265–276

Dunlap RE, McCright AM (2008) A widening gap: republican and democratic views on climate change. Environment 50:26–35

Dunlap RE, van Liere KD, Mertig AG, Jones RE (2000) New trends in measuring environmental attitudes: measuring endorsement of the new ecological paradigm: a revised NEP scale. J Soc Issues 56:425–442

Egan PJ, Mullin M (2012) Turning personal experience into political attitudes: the effect of local weather on Americans' perceptions about global warming. J Politics 74(3):796–809

Feinberg M, Willer R (2011) Apocalypse soon? Dire messages reduce belief in global warming by contradicting just-world beliefs. Psychol Sci 22:34–38

Festinger L (1957) A theory of cognitive dissonance. Row, Peterson, Evanston

Feygina I, Goldsmith R, Jost JT (2010a) System justification and environmental goal-setting: self-regulation failure as an obstacle to overcoming addiction to the status quo. In: Hassin R, Ochsner K, Trope Y (eds) Social cognitive and neuroscientific approaches to self-control. Oxford University Press, New York

Feygina I, Jost JT, Goldsmith R (2010b) System justification, the denial of global warming, and the possibility of "system-sanctioned change". Pers Soc Psychol Bull 36:326–338

Gardner GT, Stern PC (2002) Environmental problems and human behavior (2nd ed). Boston, MA: Pearson Custom Publishing

Gifford R (2011) The dragons of inaction: psychological barriers that limit climate change mitigation and adaptation. Am Psychol 66:290–302

Gupte M (2002) *Gender*, feminist *consciousness*, and the environment: exploring the "natural" connection. Women Politics 24:47–62

Hamilton LC (2008) Who cares about polar regions? Arct Antarct Alp Res 40:671–678

Heath Y, Gifford R (2006) Free-market ideology and environmental degradation: the case of belief in global climate change. Environ Behav 38:48–71

Hollander NC (2009) When not knowing allies with destructiveness: global warming and psychoanalytic ethical non-neutrality. Int J Appl Psychoanal Stud 6:1–11

Hunter LM, Hatch A, Johnson A (2004) Cross-national gender variation in environmental behaviors. Soc Sci Q 85:677–694

Jackman M (1994) The velvet glove: paternalism and conflict in gender, class, and race relations. University of California Press, Berkeley

Jost JT, Hunyady O (2002) The psychology of system justification and the palliative function of ideology. Eur Rev Soc Psychol 13:111–153

Jost JT, Hunyady O (2005) Antecedents and consequences of system-justifying ideologies. Curr Dir Psychol Sci 14:260–265

Jost JT, Kay AC (2005) Exposure to benevolent sexism and complementary gender stereotypes: consequences for specific and diffuse forms of system justification. J Pers Soc Psychol 88:498–509

Jost JT, Thompson EP (2000) Group-based dominance and opposition to equality as independent predictors of self-esteem, ethnocentrism, and social policy attitudes among African Americans and European Americans. J Exp Soc Psychol 36:209–232

Jost JT, Glaser J, Kruglanski AW, Sulloway F (2003) Political conservatism as motivated social cognition. Psychol Bull 129:339–375

Jost JT, Banaji MR, Nosek BA (2004) A decade of system justification theory: accumulated evidence of conscious and unconscious bolstering of the status quo. Political Psychol 25:881–919

Jost JT, Ledgerwood A, Hardin CD (2008a) Shared reality, system justification, and the relational basis of ideological beliefs. Soc Pers Psychol Compass 2:171–186

Jost JT, Nosek BA, Gosling SD (2008b) Ideology: its resurgence in social, personality, and political psychology. Perspect Psychol Sci 3:126–136

Kay AC, Jost JT (2003) Complementary justice: effects of "poor but happy" and "poor but honest" stereotype exemplars on system justification and implicit activation of the justice motive. J Pers Soc Psychol 85:823–837

Kendall HA, Lobao LM, Sharp JS (2006) Public concern with animal well-being: place, social structural location, and individual experience. Rural Sociol 71:399–428

Kollmuss A, Agyeman J (2002) Mind the gap: why do people act environmentally and what are the barriers to pro-environmental behavior? Environ Educ Res 8:239–260

Leiserowitz A (2006) Climate change risk perception and policy preferences. Clim Chang 77:45–72
Malka A, Krosnick JA, Langer G (2009) The association of knowledge with concern about global warming. Risk Anal 29:633–647
McCright AM (2010) The effects of gender on climate change knowledge and concern in the American public. Popul Environ 32:66–87
McCright AM, Dunlap RE (2011) Cool dudes: the denial of climate change among conservative white males in the United States. Glob Environ Chang 21:1163–1172
Moran M (2011) Psychiatry needs eyes wide open about environmental issues. Psychiatr News 46:17
Norgaard KM (2011) Living in denial: climate change, emotions, and everyday life. Massachusetts Institute of Technology Press, Cambridge, MA
Norgaard K, York R (2005) Gender equality and state environmentalism. Gend Soc 19:506–522
O'Brien LT, Major BM (2005) System-justifying beliefs and psychological well-being: the roles of group status and identity. Personal Soc Psychol Bull 31:1718–1729
O'Connor RE, Bord RJ, Fisher A (1999) Risk perceptions, general environmental beliefs, and willingness to address climate change. Risk Anal 19:461–471
Rankin L, Jost JT, Wakslak CJ (2009) System justification and the meaning of life: are the existential benefits of ideology distributed unevenly across racial groups? Soc Justice Res 22:312–333
Riechard DE, Peterson SJ (1998) Perception of environmental risk related to gender, community socioeconomic setting, age, and locus of control. J Environ Educ 30:11–19
Saad L (2007) To Americans, the risks of global warming are not imminent. Gallup Poll. Retrieved 9 Apr 2007 from http://www.galluppoll.com/content/?ci=27100&pg=1
Semenza JC, Ploubidis GB, George LA (2011) Climate change and climate variability: personal motivation for adaptation and mitigation. Environ Health 10(46)
Shepherd S, Kay AC (2012) On the perpetuation of ignorance: system dependence, system justification, and the motivated avoidance of socio-political information. J Pers Soc Psychol 102:264–280
Shiva V (1989) Staying alive: women, ecology and development. Zed Books, London
Shrivastava P (1995) Industrial/environmental crises and corporate social responsibility. J Socio-Econ 24:211–227
Sidanius J, Pratto F (1999) Social dominance: an intergroup theory of social hierarchy and oppression. Cambridge University Press, New York
Stern P, Dietz T, Kalof L (1993) Value orientations, gender, and environmental concern. Environ Behav 25:322–348
Stoll-Kleemann S, O'Riordan T, Jaeger CC (2001) The psychology of denial concerning climate mitigation measures: evidence from Swiss focus groups. Glob Environ Chang 11:107–117
Sundblad E-L, Biel A, Gärling T (2007) Cognitive and affective risk judgments related to climate change. J Environ Psychol 27:97–106
Swim JK, Clayton S, Doherty T, Gifford R, Howard G, Reser J, Stern P, Weber E (2009) Psychology and global climate change: addressing a multi-faceted phenomenon and set of challenges (A report by the American Psychological Association's task force on the interface between psychology and global climate change). American Psychological Association, Washington, DC
Swim JK, Clayton S, Howard GS (2011) Human behavioral contributions to climate change: psychological and contextual drivers. Am Psychol 66:251–264
Takacs-Santa A (2007) Barriers to environmental concern. Hum Ecol Rev 14:26–38
Tjernström E, Tietenberg T (2008) Do differences in attitudes explain differences in national climate change policies? Ecol Econ 65:315–324
Tranter B (2011) Political divisions over climate change and environmental issues in Australia. Environ Politics 20:78–96
Wakslak C, Jost JT, Tyler TR, Chen E (2007) Moral outrage mediates the dampening effect of system justification on support for redistributive social policies. Psychol Sci 18:267–274

Wang AY (1999) Gender and nature: a psychological analysis of ecofeminist theory. J App Soc Psychol 29:2410–2424

Wehrmeyer W, McNeil M (2000) Activists, pragmatists, technophiles and tree-huggers? Gender differences in employees' environmental attitudes. J Bus Ethics 28:211–222

White L Jr (1967) The historic roots of our ecological crisis. Science 55:1203–1207

Whitley BE (1999) Right-wing authoritarianism, social dominance orientation, and prejudice. J Pers Soc Psychol 77:126–134

World Bank (2009) Public attitudes toward climate change: findings from a multi-country poll. In: World development report 2010: development and climate change. World Bank, Washington, DC

Zelezny LC, Chua P, Aldrich C (2000) Elaborating on gender differences in environmentalism. J Soc Issues 56:443–457

Part V
Gender and Climate Change Examples from Around the World

Chapter 13
Gender and Climate Change in Australia and the Pacific

Margaret Alston

Abstract Climate change has emerged as one of the most significant global threats of the twenty-first century. Characterised by both incremental events such as drought and desertification, and catastrophic events such as floods, bushfires and cyclones, climate change has resulted in major social, economic and environmental consequences. To date attention has focused largely in the science of climate change and the economic consequences of declining or eroded resources. Inadequate attention to social consequences is having significant impacts on vulnerable populations. The gendered consequences of these events have only recently emerged as a major factor of international concern. This paper presents findings from work undertaken in the Murray-Darling Basin area of Australia where drought and declining water availability has had significant social and gendered consequences. It also presents findings from work underway in the Pacific region where sea level rises, temperature changes and declining infrastructure and productivity is having major social impacts and exacerbating and changing existing gender inequalities. The paper notes that in both developed and developing nations, in areas of incremental and catastrophic climate events, gender is a major factor in vulnerability, resilience and adaptation. It notes the need for gender to be a significant feature of policy and planning in relation to climate events.

Keywords Gender • Climate change • Australia • Pacific countries • Policy

Climate change is considered one of the global community's greatest twenty-first century challenges. It will critically impact on food and water security and predicted and significant global population increase will exacerbate these climate impacts

M. Alston, B.Soc., M.Litt., Ph.D. (✉)
Department of Social Work, Monash University, Dandenong Road 900, 3145
Caulfield East, VIC, Australia
e-mail: margaret.alston@monash.edu

M. Alston and K. Whittenbury (eds.), *Research, Action and Policy: Addressing the Gendered Impacts of Climate Change*, DOI 10.1007/978-94-007-5518-5_13,

creating major concerns for those most vulnerable. Across the world countries and communities will be differentially impacted. In the Gender, Leadership and Social Sustainability (GLASS) research unit at Monash University we are working on a number of projects addressing the gendered impacts of climate change. These include work in the Asia-Pacific region including Bangladesh, the Philippines, the Pacific region and Australia. We are building a body of knowledge across what is commonly referred to as the global south and are growing a centre of excellence focused on gender, climate change, women's leadership, sustainability and gender-based violence. Our expanding cohort of PhD students is allowing us to develop complex understandings of a vast geographical region in a relatively quick time.

This paper focuses on areas in the high-income nation, Australia, as well as the Pacific Island nations to its north and east to provide an understanding of the gendered impacts of climate change in our region. Australia is experiencing significant climate variability including a rise in the number of catastrophic events and has recently emerged from a 10 year drought that affected much of the continent. Climate change predictions for Australia include temperature rises, pressures on water supplies, changing cropping patterns, more frequent droughts, greater likelihood of cataclysmic events such as bushfires and floods and an increase in social and health consequences related to these events (Garnaut 2008, 2011).

In global terms perhaps the areas most vulnerable to climate changes are the Pacific Island nations that are threatened by rising sea levels, storm surges, increasing water temperatures and changes to food production. These threats will have significant social and health consequences, and are exacerbated by widespread poverty and increasing awareness that some Islands are rapidly becoming uninhabitable. The changes predicted for this region are made more complex by a growing interest on the part of the United States in the strategic value of the Asia-Pacific resulting from a more volatile geopolitical situation. The increasing economic and political strength particularly of China has shifted global power relations. This has led to a strengthening of the relationship between Australia and the US and an announcement that thousands of US troops will in future be based on Australian soil. Speaking in 2011 to the Australian Parliament, the US President Barack Obama noted that he had directed his national security team to make 'our presence and mission in the Asia-Pacific a top priority'. Further, identifying the US with the Pacific, he noted that

> As a Pacific nation the United States will play a larger and long-term role in shaping this region and its future. (Thompson 2011)

In the context of this paper it is useful to note the developing strategic importance of the Pacific as this has potential consequences for the way climate change impacts may be handled. Will attention be shaped for humanitarian and socially just solutions or for militaristic and strategic purposes, and are the two incompatible? Will attention to the region reinforce patriarchal relations thereby cementing gender inequalities and further disadvantaging women or will policies and programs adopt a gender mainstreaming approach? This chapter is designed to alert to the gendered consequences of climate events in our region and to ensure that policymakers, activists and community members do not intentionally or otherwise reinforce differential gender-power inequalities. Climate change creates an opportunity to develop strategies that address women's vulnerability, that incorporate women's knowledge and experience, that include women in decision-making and that empower women.

In a global world divided by increasing wealth and other inequalities the need to draw attention to, and address, these inequalities is paramount. There is mounting evidence across the globe that climate events and outcomes are exacerbating social inequalities and that gender is a critical factor shaping vulnerability. This chapter addresses the Australian and Pacific Island situation, noting particularly the social and health consequences of climate change and differential gendered experiences. Quite independent of climate events there is growing recognition in Australia and the Island nations of the development of uneven and highly inequitable gender power relations and increasing gender based violence (GBV). The paper draws on research conducted in Australia and the Pacific Islands to demonstrate that climate change is having gendered consequences; that women are vulnerable during and after climate events; and, conversely, that women hold critical adaptive knowledge that can aid sustainable development. Strategic attention to climate change creates an environment where gender inequalities can be challenged. Awareness of gendered vulnerabilities can also ensure that any strategic regional development incorporates attention to socially just outcomes for Australia and the Island nations and the women and men who live in these areas.

Australia and the Pacific

Australia is a large island continent in the global south with rich agricultural areas. Despite the vastness of the continent it has a small population by global standards with 22 million people, 84% of whom live within 50 km of the coast. Barely 16% occupy the vast agricultural inland areas, areas that are critical for Australia's food and water security. The fertile Murray-Darling Basin area in the eastern part of the continent covers approximately one seventh (14%) of the total area of Australia, is home to approximately 11% of Australia's population and produces approximately 45% of our agricultural produce (House of Representatives Standing Committee on Regional Australia 2011; MDB Initiative 2006). Over several years the writer has conducted research in this region noting the impacts of drought, the reduction of water in the river systems and the consequences for irrigation and broad acre farm families. One of the most critical and pressing contemporary issues is the national government's commitment to reduce the amount of water for irrigation in response to climate change predictions. This will have the effect of reducing the number of families farming and will have flow on social and health effects in the towns and communities in this vast region of the continent. It is unsurprising that this has caused significant social unrest and threatens to destabilise both federal and state governments (Alston and Whittenbury 2011).

By contrast the Pacific Island region includes twenty-two nations and territories. Nine of these island states are independent, eight are dependent territories and five are self-governing (Barnett and Campbell 2010). The Pacific Island countries//territories fall into three groups – Melanesia, Micronesia and Polynesia, and covers 30 million square kilometres of ocean (Lane and McNaught 2009), much of it, with the exception of Papua New Guinea and Fiji, being within a kilometre of the sea. By contrast with Australia, the region is home to nearly 9.5 million people of significantly diverse cultures, speaking hundreds of different languages. Per capita income is highly variable across the region and ranges from approximately $530 in Papua New Guinea to approximately $6,820 in Palau making this region not only one of the most vulnerable to climate change but also one of the most ill-equipped to deal with it.

Climate Change Impacts

In Australia climate change impacts are adding significant impetus to long-established and ongoing rural restructuring. For over half a century there have been dramatic changes taking place in rural areas as a result of technological developments, changing production practices, an increase in the size of farms in remoter areas and a growth in the number of smaller or 'hobby' farms in high amenity areas. For the last decade there has been a widespread and devastating drought across the continent, and in more recent times, severe floods and bushfires, adding to the complexity of farming. Thus climate change impacts are an added feature of existing uncertainties in agriculture. Two government commissioned reports (Garnaut 2008, 2011) on the potential impacts of climate change in Australia have detailed likely increases in temperatures, an increase in catastrophic events like drought, bushfires and floods, declining water for irrigation, changing production cycles and increasing uncertainty around food production and water use. The *Garnaut Reports* predict that there will be up to a 90% reduction in irrigated agriculture over coming decades. The second report reinforces that outcomes will be closer to worst-case predictions and suggest that by 2030 the southern part of Australia encompassing the Basin will receive 10% less rainfall and by 2050 there may be 20% less rainfall (Robertson 2010; Stokes and Howden 2010).

The Pacific Islands are particularly vulnerable to the effects of climate change because of their exposure to sea level rises, extreme weather events, rising land and sea temperatures, increased droughts, floods and extreme weather events, but also because of their tenuous livelihoods and dependence on agricultural production and fishing. Because a majority of the population lives around the coastline, this makes Island people particularly vulnerable to sea level rises and extreme weather events, events that are predicted to become more frequent. Barnett and Campbell (2009) suggest that climate impacts in the Pacific will include:

- a projected rise in temperatures including sea-surface temperature rises affecting delicate ecosystems;

- more intense rainfall/tropical storms and paradoxically less frequent storms;
- rises in sea level; and
- changes in regional climate systems which may cause droughts and food shortages.

Adaptation, Vulnerability and Resilience

Adaptation in the climate change discourse refers to actions taken to defray the effects of climate effects in order to moderate harmful impacts and enhance more beneficial outcomes (Dankelman 2010; Schipper and Burton 2009). Adapting to climate impacts will require interventions by external agencies such as governments and civil society organisations to move people and communities from short-term, and potentially unsustainable coping strategies, to more positive, sustainable adaptations. The aim of adaptation strategies is to both reduce *vulnerability* and build *resilience* amongst individuals and communities. However addressing vulnerability is dependent on understanding the political and socio-economic factors that create vulnerable groups and populations (Ribot et al. 2009). Kelly and Adjer (2009: 163) define vulnerability as the capacity to 'anticipate, cope with, resist, and recover from the impact of a natural hazard' and note that reducing vulnerability is essential for adaptation. Without this understanding, adaptation strategies may not benefit all and may in fact be harmful to various individuals and groups further exacerbating inequalities. It is therefore critical to understand gendered manifestations in order to reduce the vulnerability of women during and after climate events.

The IPCC (2008: 880) defines resilience as

> The ability of a social or ecological system to absorb disturbances while retaining the same basic structure and ways of functioning, the capacity for self-organisation, and the capacity to adapt to stress and change.

Resilience refers to the capacity of a community or individual to recover from a destabilizing event (Handmer and Dovers 2009). Handmer and Dovers (2009) describe resilience as being reactive (or resistant to change) or proactive, a process where the inevitability of change is accepted and new ways of adapting that are more sustainable are developed. It is critical to understand factors that create vulnerability within individuals and communities in order to develop adaptation strategies that build resilience.

In Australia and the Pacific nations adaptations to climate events have tended to focus on techno-scientific solutions and include the introduction of water efficiencies, attention to alternative crops and production techniques, tourism ventures and, in the Islands, the importation of much of their food requirements. At family level, one of the chief adaptations has been the adoption of alternative income generation activities including off-farm work. In Australian farm families women are most likely to be working off the farm. In Island families the sourcing of remittance income in neighbouring countries by a family member is a variant of this process of alternative income generation. Arguably there has been less intervention on the part of external agencies in Australia and the Pacific to support the need for local job creation and alternative industry development and to assist people to balance work and family expectations in order to assist adaptation and grow resilience.

Social and Health Consequences

Climate change has major global implications for food and water security. There is growing recognition that in the next decade billions of people will face food and water shortages and increased health risks as a result of climate change and that by 2030 between 28 and 67 billion US$ will be required to assist adaptation strategies (UNFCC 2007). Amongst the most vulnerable are the Island nations because of their exposure to climate events and their socio-cultural and economic status (UNFCC 2007; Barnett and Campbell 2010). There are likely to be significant social and health consequences including high mortality levels resulting from catastrophic events, declining access to fresh water and food supplies because of the inundation of productive areas and the compromising of fish habitats, increased malnutrition, increased heat stress and a rise in air and water borne diseases such as diarrhea. Sea level rises are expected to result in loss of homes and communities along coastlines and the destruction of tourism in many Island nations (UNFCC 2007; Alston and Vize 2010). The decline in food production is also impacting on availability of fresh food and this has resulted in an increase in obesity and diabetes (WHO 2010). There is evidence that water-borne diseases are affecting infant mortality in the Island nations (UNESCO 2010).

By comparison with urban areas, Australian rural areas, where climate impacts tend to be more severe, have lower adaptive capacity and the burden of climate change is higher (Molnar 2010). They also score lower on socio-economic indicators, commonly called Socio-Economic Indicators for Areas (SEIFA) including health status, life expectancy, employment, education; and income and poverty rates are much higher. Our research in the Murray-Darling Basin area of Australia suggests that there are significant social and health impacts resulting from climate variability including declining employment opportunities, educational access and income levels; physical health problems such as heat stress and an increase in mental health issues amongst farming families (Alston and Whittenbury 2011). Mental health issues arise from declining agricultural production and increased uncertainty particularly around water policy and the reduction in irrigation water. Our research indicates that social impacts in the Murray-Darling Basin include:

- increased bankruptcies;
- increasing poverty particularly amongst those associated with agriculture;
- a significant rise in levels of stress, anxiety and other health impacts;
- a widespread need for off-farm income;
- involuntary separation, particularly as women may have to leave to find work elsewhere;

- difficulty accessing employment;
- significant loss of jobs in small communities affected by drought;
- particularly men becoming socially isolated as a result of the constant need to hand feed and water livestock;
- intergenerational conflict;
- marital conflict;
- increasing feelings of alienation and mistrust;
- small community decline;
- loss of social capital evidenced in a decline in social participation;
- stress caused by water politics and a lack of information dissemination;
- difficulties in accessing income support;
- low levels of access to services;
- declining levels of individual and community resilience;
- an inability to afford farm labour, putting pressure on women and children to work on-farm;
- declining educational access amongst children because of financial problems;
- older couples continuing in farming and putting retirement plans on hold; and
- depopulation and a particular loss of young people (Alston and Whitney-Soanes 2008; Alston and Whittenbury 2010).

Our research indicates that the most vulnerable groups include those engaged in agriculture; those without the resources and capacity to 'adapt' including the unemployed; those dependent on services provided by the voluntary sector and other services that are in decline including the elderly, disabled and carers; young people whose families cannot afford the costs associated with education; and women.

Recognising the Link Between Gender, Violence and Climate Change

Meanwhile there is growing awareness that gender based violence (GBV) is increasing in Australia and the Pacific region regardless of, but exacerbated by, climate variability. Women, particularly those without the means to leave a violent situation, are particularly vulnerable to violence. Australian figures reveal that one in five women have experienced sexual violence and one in three physical violence since the age of 15 (Phillips 2006).

In 2011 a Parliamentary Roundtable on Ending Gender Based Violence in the Asia Pacific Region was held in Canberra, Australia's capital. Australian parliamentarians who are members of the Australian Parliamentary group on Population and Development (PGPD) gathered together with Members of Parliament from Papua New Guinea and Samoa, Ambassadors, High Commissioners, representatives of the international community, representatives of Australian non-government organisational community and other stakeholders (Australian Reproductive Health Alliance 2011).

The roundtable noted the link between GBV, human rights and gender equality; that GBV is directly linked to poor health; that measures to improve women's empowerment are urgently needed and that GBV is not only a human rights issue but also a *barrier to sustainable development*. This point is worth emphasizing in the context of this paper as it illuminates the link between adaptive capacity, resilience and sustainable regional development. A number of recommendations were developed from the roundtable discussion including the following that are relevant in the context of climate change.

- Women need to be heard at all levels of government. When women are involved as leaders in decision-making processes, decisions reflect women's needs and experiences;
- When communities are in stress, including from high population growth, poverty and burdened resources, GBV can be exacerbated and addressing this must include an emphasis on delivery of reproductive health services and family planning.
- Women are at greater vulnerability to climate change risks resulting from social and cultural circumstances including inadequate access to resources, displacement, and family responsibilities.
- Gender equality must be considered central in all program and service delivery.
- GBV is linked to sexual and reproductive health (SRH) and efforts to support SRH and GBV should be pursued together (Australian Reproductive Health Alliance 2011: 16–17).

There is little doubt that GBV is a growing problem in Australia and the Island nations and that this is a critical factor in women's greater vulnerability during and after climate events.

Gender and Climate Change

The impacts of climate change are rarely differentiated by gender in Australia despite emerging evidence that gender is a critical factor in vulnerability to climate events (Neumayer and Pleumper 2007; Lane and McNaught 2009). In raising the issue of gender vulnerability there is a danger in presenting women as helpless victims rather than as active agents of change. Being mindful of this, it is important to note that across the globe it is becoming clear that women are more vulnerable during and after climate events because they are more constrained by normative processes and over-represented amongst the world's poor, exacerbating their vulnerability to climate impacts (IPCC 2007).

Women own less than 15% of land and therefore are disadvantaged in decision-making forums; are over-represented amongst those most critically impacted by climate events; are 14 times more likely to die in a disaster (Neumayer and Pleumpter 2007) and are particularly vulnerable in post-disaster situations if forced to stay in shelters where they are exposed to rape, harassment, discrimination and violence; and have limited access to reproductive health services (Enarson 2009, 2006). The links between climate events//variability and violence are emerging from post-disaster sites such as Louisiana following Hurricane Katrina (Enarson 2009, 2006) and from Christchurch following the 2011 earthquake (Lynch 2011).

The increase in GBV following climate events in Australia is revealed in two recent studies. Whittenbury's (2011) research reveals that GBV has increased in the climate affected regions of the Murray-Darling Basin. This follows changes in water policy indicating that there is reduced water available for irrigation as a result of declining stocks of water. This decision has caused stress amongst irrigation families about their futures and exacerbated GBV in affected communities. Research conducted by the Women's Health Service in the Goulburn North East area of Victoria indicates that GBV increased substantially following the trauma of the Black Saturday bushfires in Victorian regions (Rout 2012).

Research in the Asia-Pacific region reveals that women are particularly at risk of sexual violence following a disaster and their vulnerability is exacerbated by a lack of access to the birth control pill. Pregnant women are at greater risk of abortion and miscarriage in shelters (Save the Children 2006) and of malaria (Mitchell et al. 2007). Poor hygiene in shelters can result in urinary tract infections, diarrhoea, skin diseases and hygiene issues related to menstruation (Ahmed et al. 2008). A particular concern is the rising level of HIV and AIDS amongst women in the Islands and this may be related to GBV and the declining status of women, particularly noted in PNG (Brouwer et al. 2007; Sepoe 2000). GBV has a major impact on adaptive capacity and resilience during and after a climate event. The likelihood of more frequent catastrophic climate events will exacerbate GBV and must be considered in policy development.

There are additional gendered impacts of climate change identified by a number of researchers. For women these include:

- an increase in women's workloads;
- higher work burdens for women in caring for the sick and injured;
- malnutrition as they prioritise food for husbands and children;
- mental health factors associated with trauma;
- loss of income as some women previously employed in tourism are now unemployed;
- inadequate support for female-headed households;
- increase in maternal and infant mortality;
- gender-insensitive reconstructive policies;
- deterioration of women's working conditions more than men's;
- slower economic recovery for women; and
- differential migration patterns (Alber 2011; Lambrou and Pianna 2006; Lambrou and Nelson 2010).

In the Island nations, because of the intense vulnerability to catastrophic climate events, there is greater recognition of the gendered impacts. Speaking in 2008 at the 52nd session of the Commission for the Status of Women in New York, the Hon Amberoti Nikora, the Kiribati Minister for Internal and Social Affairs on behalf of the Pacific Islands group, noted the need for gender to be central to climate change negotiations. He argued for actions aimed at the gendered impacts in energy, water, food security, agriculture and fisheries, biodiversity services, health, and disaster risk management and for women's traditional knowledge and practices to be part of the development of new technologies to address climate change (Lane and McNaught 2009; UNDP and AusAid 2008).

In July 2011 the writer attended a Pacific Islands Ministerial workshop in Fiji as an observer and guest of UNESCO. This forum was designed to discuss the issue of gender in the context of climate change and was attended by Ministers for Women from the various states and territories across the Pacific. The workshop attendees recognized the critical nature of climate change and the threat it poses to the vulnerable low-lying Pacific nations. They also recognized that the impacts on health and well-being and on the Pacific economies would be severe.

Delegates called for the different and complementary roles of women and men to be recognized and that different knowledge and expertise of women and men be part of climate change strategies. There is critical understanding that gender inequalities and human rights issues will hamper attempts at mitigation and adaptation to climate change and that it is therefore of critical importance to develop mechanisms for the advancement of women, that women's human and legal rights be the basis for actions, and that there be greater access to health services for women including violence services, and strategies for the economic empowerment of women to reduce their vulnerability.

The final communiqué from the workshop recognizes the need for multiple strategies to address gender and climate change including more effective participation of women in decision making around adaptation and mitigation strategies; more detailed gender analysis in relation to climate change events and policies; the incorporation of gender in strategic planning and programs; the production of sex-disaggregrated data to monitor adaptation and mitigation initiatives; further research on gender and human rights impacts of climate change; capacity building of government and civil society to mainstream gender in climate initiatives; greater valuing of the local and traditional knowledge of women and men alongside scientific knowledge to build resilient communities; climate change funding programs to be gender sensitive and benefit both women and men; strengthening collaborations and partnerships between national women's organizations and climate change departments and others organizations; incorporation of gender equality and human rights principles in regional and global negotiations; and gender and climate change to be incorporated into outcome statements resulting from regional and national meetings (Secretariat of the Pacific Community 2011: 7–8).

These communiqués from the Australian Roundtable on GBV and the Pacific gender and climate change forum recognize two complex issues – growing evidence of violence against women and that women are particularly vulnerable to climate change impacts.

Unless due attention is given to gender as a critical concept in adaptation strategies the likely impacts for women in the global south will include:

- greater likelihood of deaths and injury during natural disasters;
- higher levels of physical and mental health;
- a greater role in caring for sick and injured;
- higher levels of GBV;
- a greater role caring for sick children especially in relation to water-borne disease;
- a greater role in caring for elderly especially in relation to respiratory disease;
- greater likelihood of violence against women and breakdown of societal protections following disasters and climate events;
- disproportional risks of violence after natural disasters;
- a greater likelihood of loss of employment in tourism;
- fewer new employment opportunities;
- a greater likelihood that women will lose land rights;
- a higher rate of malnutrition as women tend to eat last;
- a greater burden of work collecting clean water;
- a heavier work burden collecting fuel;
- fewer roles in post-disaster reconstruction;
- less involvement in decision-making;
- a greater loss of status due to declining participation in post-reconstruction, higher levels of violence, and a loss of basic freedoms;
- a lack of participation in household/community decision-making;
- out-migration;
- relationship stress;
- a loss of traditional women's knowledge and this a loss of status; and
- declining socio-cultural cohesion of communities which have been disrupted/changed by climate change impacts (Alston and Vize 2010).

Conclusion

Australia and the Pacific Islands are experiencing significant climate variability causing major social dislocation and stress. At the same time the area is viewed as a strategic region in geopolitical terms. Addressing safety and stability in the region requires attention to the uncertainty created by climate events and the significant vulnerability of the region and the people whose lives are affected. There is a critical need for adaptation strategies to build individual and community resilience. However there is also a significant danger that gender will be overlooked as a critical factor in shaping responsiveness and adaptive capacity.

Barnett and Campbell (2010: 184) argue that:

> Climate change poses a new level of risk that not all [Pacific Island] communities will be able to manage. The challenge to the region is to adapt to sustain their needs and rights and values, and the challenge to the international community is to reduce emissions to the extent that such adaptation is able to be effective, and to support communities in the Pacific to adapt in ways that they see fit. This is possible and anything less is unacceptable.

I would add that it would be unacceptable to ignore factors that increase GBV and women's vulnerability during and after a climate event. At the same time ignoring women's capacity as change agents and critical adaptors would render global efforts futile. If the Pacific region is to be the focus of global geopolitical maneuvering, it is critical that policy strategies are developed that address the region's vulnerability to climate changes. In this context there is scope to address gender inequalities, to empower women and provide safety and security during times of crisis in order to build resilient people and communities and to develop positive adaptation strategies.

References

Ahmed AU, Neelormi S, Adri N, Alam S, Nuruzzaman K (2008) Climate change, gender and vulnerable groups in Bangladesh. Climate Change cell, DoE., MoEF; Component 4b, CDMP, MoFDM, Dhaka

Alber G (2011) Gender, cities and climate change: thematic report prepared for cities and climate change global report on human settlements. www.unhabitat.org/grhs/2011

Alston M, Vize S (2010) Gender and climate change in the Pacific. Gender, Leadership and Social Sustainability (GLASS) research unit, Monash University, Melbourne

Alston M, Whittenbury K (2010) The social impacts of declining water availability and ongoing drought in the Murray-Darling Basin: short report. Monash University Department of Social Work, Melbourne

Alston M, Witney-Soanes K (2008) Social impacts of drought and declining water availability in the Murray Darling basin. Institute of Land, Water and Society, Charles Sturt University, Wagga Wagga

Australian Reproductive Health Alliance (2011) Ending gender-based violence in the Asia-Pacific region: outcomes from the Australian parliamentary group on population and developments' parliamentary roundtable on ending gender based violence in the Asia-Pacific region. http://www.arha.org.au/links/33258%20ARHA%20Outcomes%20Ending%20Gender%20Based%20Violence%20in%20the%20Asia-Pacific%20Region.pdf. Accessed 24 Jan 2012

Barnett J, Campbell J (2010) Climate change and small Island states: power, knowledge and the South Pacific. Earthscan, London

Brouwer R, Atker S, Robander L, Haque E (2007) Socioeconomic vulnerability and adaptation to environmental risk: a case study of climate change and flooding in Bangladesh. Risk Anal 27(2):313–326

Dankelman I (2010) Introduction: exploring gender, environment and climate change. In: Dankelman I (ed) Gender and climate change: an introduction. Earthscan, London, pp 1–20

Enarson E (2006) Women and girls last?: averting the second post-Katrina disaster. http://understandingkatrina.ssrc.org/Enarson/. Accessed 14 Mar 2011

Enarson E (2009) Women, gender and disaster. Sage Publications India, New Delhi

Garnaut R (2008) The Garnaut climate change review final report. Cambridge University Press, Melbourne

Garnaut R (2011) Garnaut climate change review – update. Australia in the global response to climate change.http://www.garnautreview.org.au/update-2011/about-review.html. Accessed 5 Jul 2011

Handmer J, Dovers S (2009) A typology of resilience: rethinking institutions for sustainable development. In: Schipper ELF, Burton I (eds) Adaptation to climate change. Earthscan, London, pp 187–212

House of Representatives Standing Committee on Regional Australia (2011) Of drought and flooding rains: inquiry into the impact of the guide to the Murray-Darling basin plan. The Parliament of the Commonwealth of Australia, Canberra

IPCC (2007) In: Parry ML, Canziani OF, Palutikof JP, van der Linden PJ, Hanson CE (eds) Climate change 2007: impacts, adaptation and vulnerability. Contribution of working group II to the fourth assessment report of the intergovernmental panel on climate change. Cambridge University Press, Cambridge

Intergovernmental Panel on Climate Change (IPCC) (2008) Fourth Assessment Report, p. 880. http://www.ipcc.ch/publications_and_data/ar4/syr/en/annexessglossary-r-z.html. Accessed 29 Nov 2011

Kelly PM, Neil Adger W (2009) Theory and practice in assessing vulnerability to climate change and facilitating adaptation. In: Schipper ELF, Burton I (eds) Adaptation to climate change. Earthscan, London, pp 161–186

Lambrou Y, Pianna G (2006) Gender: the missing component of the response to climate change. Food and Agricultural Organisation of the United Nations (FAO), Rome

Lambrou Y, Nelson S (2010) Farmers in a changing climate: food security in Andhra Pradesh, India. Food and Agricultural Organisation of the United Nations (FAO), Rome

Lane R, McNaught R (2009) Building gendered approaches to adaptation in the Pacific. Gend Dev 17(1, March):67–80

Lynch K (2011) Spike in domestic violence after Christchurch earthquake, National. http://www.stuff.co.nz/national/christchurch-earthquake/4745720/Spike-in-domestic-violence-after-Christchurch-earthquake. Accessed 14 Mar 2011

MDB Initiative (2006) Basin statistics. http://www.mdbc.gov.au/about/basin_statistics. Accessed 2 Jul 2009

Mitchell T, Tanner T, Lussier K (2007) We know what we need: South Asian women speak out on climate change adaptation. Institute of Development Studies and ActionAid, Johannesburg

Molnar JJ (2010) Climate change and social response: livelihoods, communities and the environment. Rural Sociol 75(1):1–16

Neumayer E, Pluemper T (2007) The gendered nature of natural disasters: the impact of catastrophic events on the gender gap in life expectancy 1981–2002. http://ssm.com/abstract=874965

Phillips J (2006) Measuring domestic violence and sexual assault against women: a review of the literature and statistics Parliament of Australia, Parliamentary library. http://www.aph.gov.au/library/intguide/sp/ViolenceAgainstWomen.htm. Accessed 24 Jan 2012

Ribot JC, Najam A, Watson G (2009) Climate variation, vulnerability and sustainable development in the semi-arid tropics. In: Schipper ELF, Burton I (eds) Adaptation to climate change. Earthscan, London, pp 117–160

Robertson M (2010) Agricultural productivity in Australia and New Zealand: trends, constraints and opportunities, Australian agronomy conference proceedings, food security from sustainable agriculture. In: Dove H, Culvenor RA (eds) Proceedings of 15th agronomy conference 2010, Lincoln, New Zealand, 15–18 November 2010. http://www.regional.org.au/au/asa/2010/plenary/climate-change/7402_robertsonmj.htm. Accessed 30 Nov 2011

Rout M (2012) Disaster heartbreak breeds its own cycle of violence. In: The Australian, 28 January 2012 http://www.theaustralian.com.au/news/features/disaster-heartbreak-breeds-its-own-cycle-of-violence/story-e6frg6z6-1226255752382. Accessed 4 Feb 2012

Save the Children (2006) Watermarks: child protection during floods in Bangladesh. Save the Children UK, Bangladesh

Schipper ELF, Burton I (2009) Understanding adaptation: origins, concepts, practice and policy. In: Schipper ELF, Burton I (eds) Adaptation to climate change. Earthscan, London, pp 1–10

Secretariat of the Pacific Community (2011) Fourth Pacific women's ministerial meeting, Nadi, Republic of Fiji, 20–22 July, Communique. http://www.ifuw.org/pgwnet/docs/2011-spc-comm.pdf. Accessed 24 Jan 2012

Sepoe OV (2000) Changing gender relations in Papua New Guinea: the role of women's organisations. UBSPD, New Delhi

Stokes C, Howden SM (eds) (2010) Adapting agriculture to climate change: preparing Australian agriculture, forestry and fisheries for the future. CSIRO Publishing, Melbourne

Thompson J (2011) America 'here to stay' in the Pacific: Obama ABC news. http://www.abc.net.au/news/2011-11-17/obama-addresses-parliament/3676964. Accessed 24 Jan 2012

UNESCO (2010) EFA global monitoring report 2010 reaching the marginalized regional overview: East Asia and the Pacific, UNESCO
United Nations Development Programme (UNDP) and AusAID (2008) Workshop report: forum on the gendered dimensions of disaster risk management and adaptation to climate change. http://www.adaptationlearning.net/sites/default/files/UNDP%20PC%20Climate%20Change.pdf. Accessed 9 Sept 2010
United Nations framework Convention on Climate Change (UNFCC) (2007) Climate change: impacts, vulnerabilities and adaptation in developing countries. UNFCC. http://unfccc.int/resource/docs/publications/impacts.pdf. Accessed 24 Jan 2012
Whittenbury K (2011) Paper presented to the gender and climate change conference, Prato, Italy, September
World Health Organisation (WHO) (2010) Pacific Islanders pay a heavy price for abandoning traditional diet. http://www.who.int/bulletin/volumes/88/7/10-010710/en/index.html. Accessed 9 Sept 2010

Chapter 14
Gender Issues in Climate Change Adaptation: Farmers' Food Security in Andhra Pradesh

Yianna Lambrou and Sibyl Nelson

Abstract This work contributes to a growing body of literature on gender and climate change (e.g. Lambrou and Piana, Gender: the missing component in the response to climate change. Food and Agriculture Organization of the UN, Rome, 2005; Ray-Bennett, Environ Hazards, 8(1): 5, 2009). Gender is one of numerous important socio-cultural dimensions typically included in climate change vulnerability assessments but it is rarely incorporated in adaptation research and planning. An innovative methodological model for studying gender and climate variability for use in the context of climate change was developed. Gender-sensitive qualitative and quantitative methods and gender analysis techniques were used to capture the voices of both men and women and quantify the degree to which men's and women's responses to climate variability differ. This research tests the hypothesis that due to gender roles men and women deal and cope with climate variability differently. The findings confirm that there is a strong gender dimension to the way in which climate variability is experienced and expressed by farmers in their coping strategies to ensure their livelihoods and food security. Women's and men's perceptions of and responses to impacts of changed climatic conditions differ in important ways as they pursue food security. Such findings are essential for informing policy decisions by ensuring that the experiences of both women and men are embedded into policy design.

Keywords Gender • Food security • Climate change • India • Agriculture • Livelihoods • Adaptation • Methodology • Coping strategies

Y. Lambrou (✉) • S. Nelson
Food and Agriculture Organization of the UN, Viale delle Terme di Caracalla, Rome 00153, Italy
e-mail: Yianna.lambrou5@gmail.com

M. Alston and K. Whittenbury (eds.), *Research, Action and Policy: Addressing the Gendered Impacts of Climate Change*, DOI 10.1007/978-94-007-5518-5_14,

Introduction

The aim of this research was to compare how recorded climate trends correspond to men and women farmers' perceptions and to understand how men and women in farm households perceive and experience climatic shifts and how this is linked to food security. Further, the study aimed to identify the coping strategies that men and women farmers utilize in order to ensure a measure of food security in response to climate variability as well as the resources, decision making processes and institutional support utilised. Finally, an overarching aim was to develop a replicable methodology for examining the gender dimensions of farmer responses to climatic variability and change.

Background

Development in Andhra Pradesh

Of the 83 million inhabitants of Andhra Pradesh, around 60–70% rely on agriculture as their primary livelihood (Government of Andhra Pradesh 2008; Acosta-Michlik et al. 2005). Despite advancements in the socio-economic development of Andhra Pradesh overall,[2] such as having one of the lowest levels of poverty in India at around 15%, farmers in the rainfed agriculture zones of Andhra Pradesh continue to face development and environmental challenges to their livelihoods, food security, and overall wellbeing (Government of Andhra Pradesh 2008), including limited employment opportunities, low incomes, environmental degradation, and social inequalities.

These challenges are particularly daunting in the study districts: Mahbubnagar (in the Telangana region) and Anantapur (in the Rayalaseema region). According to the *2007 Human Development Report – Andhra Pradesh* (Government of Andhra Pradesh 2008), both districts are among the poorest in the state, and rank near the bottom of the 23 districts of Andhra Pradesh according to the Human Poverty Index.[3] The *Human Development Report* also found that the districts perform poorly with regard to gender equality and women's empowerment, with Anantapur ranking near the bottom and Mahabubnagar in last place among the districts of Andhra Pradesh, according to its measurements.

Climate Variability in Study Area

Both Mahbubnagar and Anantapur have a history of drought with both districts experiencing drought during approximately 20% of the 40 years between 1960 and 1996 (Acosta-Michlik et al. 2005) largely linked to low levels of agricultural

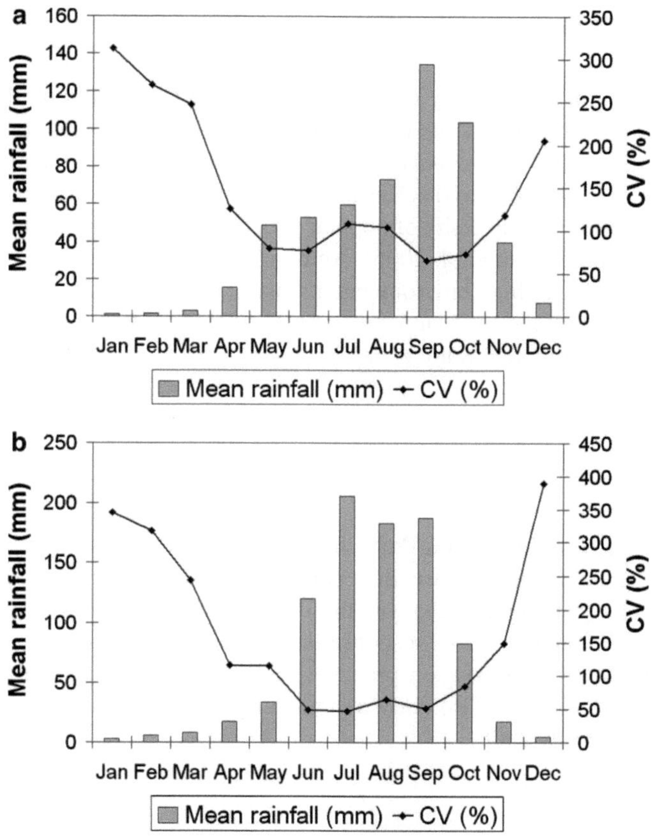

Fig. 14.1 Monthly mean rainfall (mm) and coefficient of variation (CV) in (**a**) Anantapur district and (**b**) Mahbubnagar district

output (World Bank 2006). Poor farming households are the most vulnerable and in times of drought, those who are unable to change their farming practices may face starvation, loss of health and loss of life (World Bank 2006). Under the difficult conditions of farming in these drought-prone areas, smallholder farmers must diversify their livelihoods (Deb et al. 2002). A common strategy is to work as a day labourer or migrate for labour.

Anantapur is one of the most drought-prone districts in Andhra Pradesh. The annual average rainfall is 552 mm, out of which more than 60% is received between June and September during summer monsoon season. Mahbubnagar receives an average annual rainfall of 850 mm, with monthly rainfall showing a dominant uni-modal monsoon season pattern with a maximum rainfall in July. The co-efficient of variation of mean monthly rainfall of Anantapur and Mahbubnagar shows a high level of variability especially during crucial crop growing months. Note that July rainfall is crucial for agriculture as most of the rainfed crops are sown and rice transplanting also takes place during this month. See Fig. 14.1.

Climate Change

Global climate change is likely to pose new stressors to agricultural development and farmers' livelihood. Despite uncertainties in the projections of climate change impacts, it appears that the increasing temperatures associated with climate change, combined with other pressures such as increasing population, decline in soil fertility, and decrease in genetic diversity of popular varieties, may lead to a decrease in crop production in India (Mall et al. 2006). Andhra Pradesh is considered vulnerable to climate change due to high environmental sensitivity, including the amount of safe water, percentage of managed land compared to total land, and ratio of fertilizer use to land area (Brenkert and Malone 2005).

Adaptation

For decades farmers in Andhra Pradesh have relied on coping strategies in order to get through unfavourable weather conditions, particularly drought. With the added pressures of climate change, these coping strategies may no longer be enough. Adaptation is a crucial response in the agricultural sector and in the study area includes water conservation practices, enhancement of existing anti-drought programs (World Bank 2006) and strengthening communities' capacity to adapt (Prabhakar and Shaw 2008).

Methodology

The study explores how coping strategies are shaped by the gender dimensions of power relations, access to resources and equity in decision making. By demonstrating men's and women's abilities to react to changes in the climate, the findings show us what is needed for longer term climate change adaptation.

Study Area

The research was carried out in six villages in two drought-prone districts of Andhra Pradesh, Anantapur and Mahbubnagar. Agriculture in both districts has shifted from the cultivation of a relatively high number of food crops for household consumption (sorghum, pigeon pea, castor, minor millets, horsegram) to fewer, newer crops for securing income for household expenditure, primarily groundnut.[4] Seasonal rainfall has a significant influence on the yield of the crops.

Borewells were introduced in all of the study villages in the 1980s, along with increased mechanization and a shift toward fertilizer and pesticide use. In recent years the water table has declined significantly so many of the existing borewells are now of little use.

Participants

The participants in the study were drawn from the same portion of the local population. They were farmers with land holdings below 5 acres, relying primarily on rainfall, between 40 and 69 years old living in a household, with at least 10 years engagement in farming.

Almost all of the respondents were from households that owned rainfed land (95%), while some respondents, 42%, owned some irrigated land in addition to rainfed land. The majority of participants had little formal education and women were very likely to be illiterate. Almost all of the respondents were Hindu and the majority belonged to the caste traditionally composed of artisans and farm labourers ("the Backward Caste").

Data Sources and Analytical Methods

An analysis of climate indicators for the key crops in the region to characterize the climate variability and trends of the past four decades was conducted using daily temperature and rainfall data for a representative station in Mahbubnagar and one in Anantapur.

Focus group discussions and key informant interviews were used to collect qualitative data on men's and women's perceptions of environmental change, livelihood options and food security.

Male and female heads of household from 95 households each responded to a quantitative survey,[5] for a total of 190 individual respondents. Views from members of the same household were collected to control for variability between households and to isolate the gender dimension. In addition, 11 female heads of households took part.[6] The data from the questionnaires was entered into the Statistical Package for Social Sciences (SPSS). Cross-tabulation analysis was performed to examine gender differences in perceptions of changes in climate and related food security, as well as coping strategies and institutional support. Statistical differences and significances were obtained using chi-square.

Research Results & Analysis

Climate Variability

Men and women farmers in both study districts reported similar perceptions of climate variability and long-term trends. Furthermore, the farmers' perceptions were similar to the record of climate trends over the past 40 years.

All farmers participating in the quantitative survey agreed that the weather had changed over the past 30 years, and the recorded data suggests a shift in average climate conditions. Seventy eight percent of respondents (81% of men and 75% of women) agreed that the temperature had increased while 97% of respondents (97% of men and 97% of women) agreed that the amount of rain was less. When asked why these changes were happening, most respondents said they do not know or "only God knows". While it might not be clear to the farmers why the changes were happening, it is clear from the recorded data and farmers' perceptions that the conditions with which farmers are familiar were changing and their livelihoods and well-being appeared to be under stress.

Farming Activities Impacted by Changes in Weather[7]

Based on the changes in weather they had observed, the participants were asked to describe what those changes meant for their lives. Men and women speaking in separate focus group discussions recognised the centrality of rainfall for their livelihoods and said that "no rains" or "no timely and sufficient rains" were the primary risks facing agricultural production. The survey results also revealed that men and women had similar perceptions of the impacts of the changes in the weather. Ninety percent of men and women surveyed reported that weather changes had led to poorer harvest or reduced crop yields (Table 14.1).

Livelihood Activities Impacted by Changes in Weather

Significant gender differences existed in who was perceived as most adversely affected by changes in the weather. Both men and women agreed in similar numbers that the entire family was most affected by the changes. However, women were much more likely than men to report that women were most affected. At the same time, men were more likely to report that men were most affected (Table 14.2).

The four most popular responses to the question "how are **men's** lives today different than 30 years ago because of the changes in the weather?" were:

Table 14.1 Perceived impact of changes in weather over the past 30 years on key components of farming activities, by gender

Impact of changes	Percent of male respondents reporting change	Percent of female respondents reporting change	Percent of all respondents reporting change	X^2
Less fodder	45.3	28.3	38.3	6.23**
Use different crops	15.8	15.1	15.4	.019
Borewells/ponds dry up	53.2	46.8	55.2	3.45*
Decreased livestock	42.1	36.8	39.3	.593
Shortage of drinking water	32.6	26.4	29.4	.934
Poorer harvest/yield	87.4	88.7	88.1	.082
Reduced forest	16.8	18.9	17.9	.140
Health affected	17.9	28.3	23.4	3.02*
Household Garden losses	1.1	1.9	1.5	.237

*X^2 sig @.10
**X^2 sig @.01

Table 14.2 Perception of who in the family is most affected by changes in the weather in the past 30 years, by gender

Who is affected	Percent of male respondents	Percent of female respondents	Percent of all respondents
Men	23.2	21.7	22.4
Women	7.4	26.4	17.4
Children	1.1	0.9	1.0
Elderly	0	0.9	0.5
Entire family	67.4	49.1	55.7

1. Increase in pressure to mobilize loans (reported by 61% of all respondents)
2. Increase in pressure to provide food (reported by 51% of all respondents)
3. Increase in emotional stress/anxiety (reported by 33% of all respondents)
4. Increase in fights/arguments among the family (reported by 22% of all respondents)

Conversely, the four most popular responses to the question "how are **women's** lives today different than 30 years ago because of the changes in the weather?" were:

1. Increase in pressure to provide food (reported by 61% of all respondents)
2. Increase in workload at home (reported by 55% of all respondents)
3. Increase in health problems (reported by 36% of all respondents)
4. Increase in fights/arguments among the family (reported by 28% of all respondents)

These results show that changes in climate over time have different impacts on men and women and these differences are linked to their gender roles.[8] The result that an increase in emotional stress/anxiety was a highly-ranked issue for men more so than women is somewhat surprising as women would be expected to be more likely than men to report an increase in emotional stress. The increase in emotional stress among men was likely due to the pressure to mobilize loans, which the respondents report as the top pressure and which has been documented elsewhere as a serious issue, sometimes leading male farmers to commit suicide (Sainath 2004).

When a distinction is made between how men and women view the effect of weather changes on men's lives, it became clear that there was a gap between men's and women's perceptions about men's lives. This difference in men's and women's perspectives was also true when considering changes in women's lives.

These results demonstrate the importance of documenting men's and women's views of their own lives so that the degree to which certain issues matter more to men or women is clear. It is important to collect data from men and women separately so that their responses are not biased by the presence of members of the opposite sex (e.g. men may be less likely to report on emotional issues in front of women if cultural definitions of masculinity prohibit displays of emotion).

Perception of Climate Impacts on Food Security

Men and women farmers both reported an increase in pressure to provide food to the family as one of the primary changes in their lives due to changes in the weather. Other areas related to food security, however, did not rank as highly.

In contrast, during qualitative discussions, most farmers did mention a decrease in quality of food over the past 30 years. Based on their responses to the quantitative survey, it appears that farmers may not have associated the decrease in quality with changes in the weather. Instead, they may have perceived the decreasing quality of food as resulting from the shift toward purchasing food (even though this shift is partly related to climatic changes affecting the viability of farming). The issues of availability and sources of food are key components of food security in relation to a changing climate and are discussed in subsequent sections of the report.

While the farmers were asked to list changes in their lives due to changes in the weather over the last 30 *years*, it appears that some of their responses were not changes due only to weather. Over the course of the last three decades, many changes have taken place in farming practices, infrastructure and government support which have intersected with changes in climate variability. It is not possible at this level of analysis to attribute changes in farmers' lives directly to or solely to changes in climate variability over time. The important outcome here is that men and women farmers perceived changes in their well-being linked to a changing climate and these changes differed for men and women. The following section examines how farmers responded to weather conditions in a given year in the context of climate variability; on this time scale, causal linkages are more clearly defined.

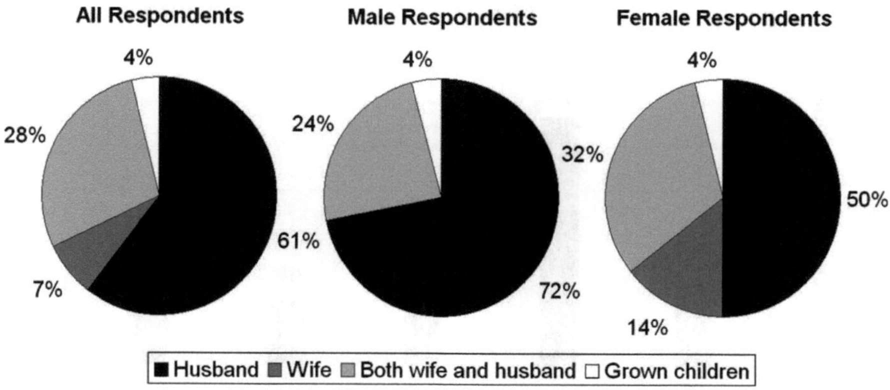

Fig. 14.2 Who decides to grow different crops or to change cropping patterns? Comparison of responses by men and women

Coping Strategies

During 2008[9] (January-December), Mahbubnagar experienced below normal rainfall during the growing season, while Anantapur experienced above normal rainfall during that time. Farmers' perceptions confirmed these patterns.

In response to the low and late-arriving rainfall of 2008, the majority of the Mahbubnagar respondents made some sort of change or a combination of changes to their cropping and livestock practices, the most frequent being cultivating crops that used less water (50%) and changing to more edible crops (37%). Indeed, rainfall was the primary factor among almost all respondents (96%) in influencing changes in cropping patterns. In a year in which there was *more* rain than 2008 (i.e. normal rainfall), farmers in Mahbubnagar would change their cropping pattern (48%) and grow more kinds of crops (39%). Similar percentages of men and women responded with this preference.

Meanwhile, in Anantapur, in response to an above-normal amount of rainfall, 63% of farmers made no change to their cropping and livestock practices while 25% did change their cropping pattern. This could be indicative of the respondents not having the resources to make changes, not having the information to anticipate the above-normal rainfall or not believing changes were necessary. Farmers did seem to feel that they could implement changes if necessary; most farmers in Anantapur said that in a year in which there was less rainfall than 2008, they would cultivate crops that required less water.

There was no discernible difference between men and women with regard to what they reported as changes made in response to abnormal rainfall in Mahbubnagar and Anantapur. However, men and women did have distinct roles in deciding to implement changes and they also had different perspectives on their roles. See Fig. 14.2.

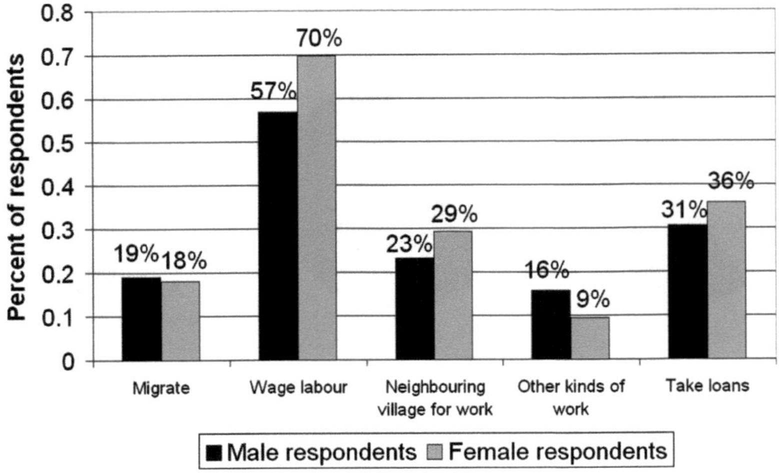

Fig. 14.3 Men's and women's coping strategies to earn income when there is a crop loss due to low rainfall

Choice of Wage Labour[10]

In general, when rainfall led to low yields and insufficient income, as it did in 2008 in Mahbubnagar and in other years in Anantapur, farmers employed various coping strategies to supplement their farm income. As shown in Fig. 14.3, the majority of farmers turned to wage labour for income, for example, in construction. More women (70%) than men (57%) reported that they undertook wage labour as a coping strategy in response to crop loss, which could be indicative of women's roles as "farmers" being less socially determined in terms of roles and expectations than men's. In addition, women of lower castes took on wage labour more readily than men; women of lower castes supplementing family income through wage labour has also been shown elsewhere (Deb et al. 2002). Taking loans followed by taking work in neighbouring villages were also mentioned by about one third of participants, in both cases, by a higher percentage of women than men.

The majority of respondents, both men and women, cite migration for income (as opposed to migration for survival, see Deshingkar and Start 2003; Rao 2001) as the main coping strategy in response to drought. In response to drought, local wage labour and taking loans were also employed as coping strategies. The question of who migrates seemed to be related to the severity of the drought. In focus group discussions, farmers spoke of large numbers of whole families migrating in response to severe droughts in the 1970s and many remaining away permanently, whereas today families migrated to cities for construction work during the non-agricultural season but not permanently. In some cases the husband migrated and the rest of the family stayed behind. In the study villages, farmers reported in qualitative exercises that the decision whether to migrate and who migrates were taken jointly by men and women, however the specific decision of where to migrate was taken by men.

Other studies in Andhra Pradesh have shown that migration in search of wage labour is a major coping strategy of households affected by drought (World Bank 2006; Laxmaiah and Vijayaraghavan in Rao et al. 2005). Some studies have documented the hazardous conditions that migrants encounter, including exploitation of labourers from Mahbubnagar district by labour contractors (Olsen and Ramana Murthy 2000) and the employment of women and children from Anantapur in waste collection in Hyderabad whose living conditions are without basic amenities (CARPED 2003). This demonstrates that farmers in this region no longer relied only on farming and in drought years, and both men and women were likely to seek employment elsewhere.

If confronted with persistent below-normal rainfall or dry conditions in the future, farmers' coping strategies would likely shift from modifying on-farm activities to other income-generating opportunities. If there were not enough rain for a few years in a row, respondents suggested their preferences would be as follows: (1) taking waged labour (59%): (2) going to neighbouring villages for work (35%): (3) migrating (27%): (4) and taking loans (27%). There was no significant gender differentiation in these responses.

However when asked if the weather was no longer predictable from year to year, i.e. not just persistently dry as above but continually unpredictable, gender differences emerged. Both men and women would seek additional income through wage labour, however women would prefer to do so closer to home whereas men were prepared to go farther away. More male respondents preferred migration in search of wage labour (47%) to local wage labour (38%) as a coping strategy, whereas more women preferred local wage labour (57%) to migration in search of wage labour (18%) as a coping strategy. This statistically significant difference along gender lines suggests that if the climate becomes unpredictable for the foreseeable future, men were prepared to go further away. There are implications of these different preferences for family structure and their long term sustainability, and how they will be reconciled with each other.

Coping with Food Insecurity

Ninety-eight percent of respondents reported that the amount of income from farm produce, which was their primary source of income, followed by wage labour, varied from year to year and was linked to climate variability. Income from farm produce was impacted in order of importance by the weather, yield, prices, availability of power supply and labour availability. This suggests that farmers' food security varies from year to year along with income.

Income is critical to the food security of the farming households, as food is sourced primarily through purchases rather than only through their own production. In a year with below normal rainfall, farmers obtained their food from multiple sources, including the Public Distribution System (PDS) (94%), the open market (75%), stored produce (53%), through earnings of wage labour (48%) and wife's original home (in a patrilineal society the wife moves to the husband's house but continues to be assisted by her family) (3%). Less than a third of respondents noted they could not buy sufficient food for the entire family, suggesting that for the majority of respondents, sufficient amounts of food were available, despite the unfavourable climate conditions. While there was *availability* of food, the *utilization* component of food security did not appear to be sufficient in a dry year. One quarter of respondents reported that their households had sufficient food but not the types of food they wanted to eat. Most respondents found the food to be somewhat nutritious and a third did not think it was at all nutritious.

For those families that did not have sufficient food during a low rainfall year, the coping strategy for dealing with this shortfall was for all family members to eat less. However, more women than men noted that they themselves would eat less. In fact, men were more likely than women to say that the whole family had sufficient food (56% of men vs. 34% of women). Also, women were much more likely to say that their husband got sufficient food (24%) than men would say their wives got sufficient food (1%). As men were not involved in food preparation they were more likely to be unaware of the actual food distribution. In the qualitative focus group discussions, women described eating two instead of three meals: they distributed food first to men, then to children (boys and girls) and finally to themselves. This gender-based distribution of food is a traditional coping strategy for dealing with food scarcity and takes place usually even without low rainfall (and is based on the 'value' and 'worth' culturally assigned to men and women); however, it appeared to be exacerbated during a dry year. In-household *access* to food must be considered when examining how climate variability impacts food security.

Finally, the *stability* of the farmers' food security appeared to vary over the course of the year and was tied to climate variability. The men and women of Malkapur village (Mahbubnagar district) described how during the June to October season they relied on the Public Distribution System (PDS)[11] for mostly white rice and stored food from their own crops. According to the women, the amount of food available largely depended on the men's decision on how much to store and how much to sell. Women also noted that while it was the men's responsibility to sell the crop, women, who were responsible for ensuring the family's food supplies and also for cooking and distribution of food, would hide some of the crop without the knowledge of the men for the family's consumption.

In Malkapur, the rains in the June to October season were late and yields were low, meaning income was also low. The farmers sought income from other sources in order to pay the loans they had taken for farming activities during June to October and to prepare the fields for the November to February season. The yields during the November to February season were also low, and so by the start of the third season, some of the men of Malkapur migrated and the women collected firewood to sell in order to overcome their severe food shortages.

Access to Institutional Support

Income Support Options

Indebtedness[12] among small farmers in Andhra Pradesh has been documented as a major problem with multiple causes, including agricultural stagnation, increased production and marketing risks, an institutional vacuum, a lack of alternative livelihood opportunities, and environmental degradation (Government of India 2007).

In reference to the year 2008, 53% of respondents noted they took loans, although this differs significantly by gender with 66% of men and 41% of women reporting they had taken loans.[13] The source of loans varied by gender, but the main source was the bank, followed by self help groups and big farmers or money lenders. The amount of the loans ranged from Rupees 3,000 to Rupees 930,000, with the majority in the range of 10,000–59,000. Nine of the eleven female-headed households in the study population received loans, ranging from Rupees 4,000 to 40,000 (average Rupees 24,000). These figures suggest that while women have access to loans, they may have access to smaller loans than men, impacting their coping options.

Information for Farming and Adapting

Much less than one quarter of respondents reported receiving assistance (information, trainings, materials) from the government agricultural department, although this is significantly gendered with 36% of men mentioning the agricultural department and only 22% of women mentioning it. The type of support received was also gendered: 23% of men and 7.5% of women reported receiving agricultural information. In qualitative focus groups, farmers reported that the extension networks were poor and considered the advice not to be appropriate.

With regards to receiving information related to on-farm production, 33% of respondents said they receive information on cropping patterns/practices, however this was significantly gendered with 47% of men and 21% of women responding positively. Most agreed that more information was given to farmers with larger land holdings (who happened to always be male); there was a general perception that women farmers did not receive information (this was not necessarily related to the size of land holding but primarily due to gender) although this was not confirmed in the quantitative analysis. The expectations of government assistance also appeared to be gendered; more men than women reported knowing about and receiving timely information on weather and irrigation schedules; whereas women were less aware of the availability of such assistance.

Men and women appeared to have different sources of information about the weather. Television and relationships with neighbours in obtaining information related to weather forecasts were two important sources. In particular, women appeared more likely than men to rely on neighbours for information, whereas men appeared more likely than women to rely on their traditional knowledge. These sources of information were important factors in disseminating information to farmers on responding to weather events and seasonal climate conditions.

The results of gender differences in access to institutions and information reflected differences in men's and women's education levels and literacy, as well as their culturally-defined roles in decision making and division of labour. Due to cultural practices gender was a factor in accessing precisely the information and support necessary for responding to climate variability, suggesting that women are disadvantaged with regards to access to institutional support, a fact with major implications for building resilience to long-term climate change.

Discussion

Relevance of Findings for Climate Change Adaptation

Although the findings presented here relate to climate variability, they are relevant for adaptation to long-term climate change because they illuminate how men and women do and do not cope with changes in their livelihoods and food security induced by climatic factors.

Looking to the future of the farmers in the study area, as well as broader implications outside this specific context in other parts of India, it is clear that there are multiple trends – global, national and regional – that threaten food security and rural well-being in the long term. These agricultural regions are increasingly reliant on food being brought in from outside as subsistence production declines, raising the question of how future food security will be ensured. The farmers have come to rely on loans in order to carry out their farming activities, but are often unable to pay these loans and it is unclear how the trap of indebtedness can be resolved. The livelihood choices farmers make are embedded in socio-economic as well as cultural traditions and practices which are also changing at various speeds. Farmer decision-making appears to take place most often on a short and less so on a long-term basis as farmers manage multiple and sometimes contradicting factors. This may call into question their capacity to plan for longer-term shocks to their livelihoods but it may also signal new opportunities for change and social renewal if the right support and capacity building is provided.

Given these trends as well as well as changes in climate, how will men and women farmers fare? In particular, how will men's and women's roles in decision-making and their access to resources shape the future of food security and livelihood activities? Furthermore, how might gender roles and relations change along with climate change? As women become increasingly responsible for the household income when the husband migrates, will this signal the greater empowerment and agency of women? It would appear that as their social roles shift, women may gain confidence in dealing with other socio-cultural inequities they face daily. However, without external resources for women to become full-fledged farmers, their empowerment remains precarious and open to challenge.

Incorporating Gender Issues into Climate Change Adaptation

1. Climate, crop and rainfall information dissemination

In the future it will be essential to have a system of information that is accessible widely and is applied without bias due to gender, property ownership, caste, age or religion. It is of crucial importance that institutional information is received by women as well as men through advice on radio and television in the form and content that is appropriate, and through legitimate community information sources as well as through women's organisations. This information should be tailored to the decision-making needs of farmers and must arrive in a timely fashion.

2. Methodology

More research based on gender analysis that recognises the heterogeneous nature of men and women farmers and the variety of strategies that they develop and employ continuously is needed to enhance policy recommendations. Further cross-regional intra-household analysis would illuminate the inequalities of the approaches in use and impacts for different members due to gender, age and access to resources.

3. Food security

Given the changes in livelihood activities that are taking place in part in response to changes in climate, it appears that future food security will probably be much more determined by the type and location of employment opportunities (which determine access to food) than by the availability of food through on-farm production. Greater attention needs to be paid to the changing situation of farming in India (and elsewhere) to ensure food security in both rural and urban areas on the whole. Leaving farming for government employment programmes or outright migration may assist the household temporarily but may change permanently the role agriculture can play in development.

4. Short- versus long-term planning

Current coping strategies must be weighed against shorter term adaptation with a view to longer-term sustainability. As one survival or coping strategy, eating less may be helpful in assuring the overall existence of the household but it puts women at great personal risk and vulnerability. Using increasingly scarce resources such as water (from borewells) may assist the households to survive for the moment, but may be destroying forever the un-replenished aquifers of the future. Criteria to evaluate (for sustainability and efficiency) the documented coping strategies on a variety of levels from the personal to the household, community, regional, national and international will need to be applied. Otherwise, short term solutions and strategies might be detrimental to the available natural resources undermining the future capabilities of farmers to cope with crises in the long term.

5. Participation

Farmers' risk assessments may be different from official accounts and should be heard. Based on the study findings, it is legitimate to foresee that some people will try to leave agriculture and some lands will be abandoned. It would be important to invest in increasing production efficiency, and technical support such as farmer educational programmes and advisory services addressing both men and women's roles and needs. At the same time, changes in access at the household and individual level would also need to be addressed.

6. Future resilience

It is increasingly impossible to continue 'business-as-usual' in agriculture and the fact that external threats demand new solutions, opportunities emerge for women and men to participate in untried decision-making processes as well as to take on newly defined leadership roles. Gender relationships are also under flux (particularly within younger couples) and this creates new spaces for social change and the emergence of new social patterns and opportunities.

By strengthening emerging social openings for socio-cultural change, new role models may lead to the creation of longer-term sustainability options. The differences documented here by gender in perception of food security show the importance of consulting both men and women when seeking to understand how food security may be related to climate shifts both short and long-term. The research reveals that gender does matter when assessing farmers' responses to climate variability and long-term changes in the climate.

Conclusions

Planning for adaptation to long-term change must be founded on men and women farmers' specific knowledge and experiences as they make choices in an uncertain climate. In addition, future plans and government policies must consider the reality of farming today, including migration and what the implications will be for longer-term community and family stability and the role of agriculture in development. In pursuing strategies for the future, the needs of women and men farmers need to be incorporated in regional, national and international development plans to ensure an integrated and gender-sensitive approach based on sound knowledge and strategies that build the resilience of the most vulnerable to the impacts of climate change and the enhancing role that agriculture can play in ensuring food security.

Notes

1. This paper is drawn from a report produced at the Food and Agriculture Organization of the United Nations (FAO). Views expressed are those of the authors and do not necessarily reflect the views of FAO.
2. Covered in detail in Government of Andhra Pradesh 2008 and Acosta-Michlik et al. 2005.
3. The Human Poverty Index measures the distribution of progress in achieving human development through the level of deprivation. Anantapur receives a score of 0.515 (ranking 20 out of 23 districts) and Mahbubnagar receives a score of 0.592 (ranking 22 out of 23) using 2001 figures.
4. The spread of the market economy and government subsidized grains in the region reduced the need for cultivating food crops for home consumption. In addition, oil mills led to an increase in the price for groundnut and the groundnut based cropping system became profitable for farmers (Gadgil et al. 2002).
5. The surveys were administered to men and women separately so that they did not influence each other's responses.
6. While this article focuses on the gender roles and relations of husbands and wives, the primary decision makers in the study households, it is noted that gender roles are more complex and are shaped over time by age, caste and class.
7. The term "changes in weather" here can be considered as interchangeable with "changes in climate variability".
8. It was not possible within the scope of the study to analyse responses based on age or to analyse in detail the outcomes for different age groups.
9. The year 2008 was chosen as it was the most recent full year, so it is likely that farmers' memories of their actions are accurate.
10. The most common coping strategies mentioned by the farmers are discussed.
11. According to the Department of Food and Public Distribution (2010), the Public Distribution System (PDS) "evolved as a major instrument of the Government's economic policy for ensuring availability of foodgrains to the public at affordable prices as well as for enhancing the food security for the poor". The PDS is operated jointly through the Central and State Governments. Under the PDS, wheat, rice, sugar and kerosene are distributed via Fair Price Shops.
12. For more on indebtedness in Andhra Pradesh and the link between the debt trap and farmer suicides, see Government of India 2007.
13. This difference is in line with gender roles in the sphere of financial matters in which men play a dominant role. 51% of women noted their husbands had taken loans and 22% of men that their wives had taken loans.

Acknowledgements Thank you to Francesca Mancini, Selvaraju Ramasamy, Daniel Gustafson, Gavin Wall, Marcela Villarreal, Caroline Dookie, Reuben Sessa, Claudia Escutia and Sonia Zepeda. In addition, the authors thank Nandini Prasad, Carolyn Sachs, Margaret Alston, M. Bharath Bhushan, Lisa Schipper, Prasanthi and colleagues, P. Raghava Reddy, V. Adinarayana, D. Raji Reddy and Atsuko Nonoguchi. The authors acknowledge with appreciation the grant from SIDA.

Literature Cited

Acosta-Michlik L, Galli F, Klein RJT, Campe S, Kumar K, Eierdanz F, Alcamo J, Kromker D, Carius A, Tanzler D (2005) How vulnerable is India to climatic stress? Measuring vulnerability to drought using the security diagram concept. In: Human security and climate change – an international workshop, GECHS, UNEP, IHDP, CICERO, CSCW, Oslo

Brenkert AL, Malone EL (2005) Modeling vulnerability and resilience to climate change: a case study of India and Indian states. Clim Chang 72(1–2):57–102

Centre for Action Research and People's Development (CARPED) (2003) District profiles of Andhra Pradesh. CARPED, Hyderabad

Deb UK, Rao GD, Rao YM, Slater R (2002) Diversification and livelihood options: a study of two villages in Andhra Pradesh, India 1975–2001. Overseas Development Institute, London

Department of Food and Public Distribution (2010) Department website. http://fcamin.nic.in/dfpd_html/index.asp

Deshingkar P, Start D (2003) Seasonal migration for livelihoods in India: coping, accumulation and exclusion (working paper 220). Overseas Development Institute, London

Gadgil S, Rao PS, Rao KN (2002) Use of climate information for farm-level decision-making: rainfed groundnut in southern India. Agric Syst 74(3):431–457

Government of Andhra Pradesh (2008) Human development report 2007 – Andhra Pradesh. Centre for Economic and Social Studies and Government of Andhra Pradesh, Hyderabad

Government of India (2007) Report of the expert group on agricultural indebtedness. Ministry of Finance, Government of India, no city

Lambrou Y, Piana G (2005) Gender: the missing component in the response to climate change. Food and Agriculture Organization of the UN, Rome

Lambrou Y, Nelson S (2010) Farmers in a changing climate: food security in Andhra Pradesh, India. Food and Agricultural Organisation of the United Nations (FAO), Rome

Mall RK, Singh R, Gupta A, Srinivasan G, Rathore LS (2006) Impact of climate change on Indian agriculture: a review. Clim Chang 78:445–478

Olsen WK, Ramana Murthy RV (2000) Contract labour and bondage in Andhra Pradesh (India). J Soc Political Thought 1:2

Prabhakar SV, Shaw R (2008) Climate change adaptation implications for drought risk mitigation: a perspective for India. Clim Chang 88(2):113–130

Rao GB (2001) Household coping/survival strategies in drought-prone regions: a case study of Anantapur district, Andhra Pradesh, India. SPWD-Hyderabad Centre, Hyderabad

Rao KPC, Bantilan MCS, Singh K, Subrahmanyam S, Deshingkar P, Rao PP, Shiferew B (2005) Overcoming poverty in rural India: focus on rainfed semi-arid tropics. International Crops Research Institute for the Semi-Arid Tropics, Patancheru

Ray-Bennett NS (2009) The influence of caste, class and gender in surviving multiple disasters: a case study from Orissa, India. Environ Hazards 8(1):5

Sainath P (2004) Farmers' suicides in Andhra *India Together*. http://www.indiatogether.org/opinions/psainath/suiseries.htm. Accessed 1 July 2010

World Bank (2006) Overcoming drought: adaptation strategies for Andhra Pradesh, India. World Bank, Washington, DC

Chapter 15
Climate Change, Women's Health, Wellbeing and Experiences of Gender Based Violence in Australia

Kerri Whittenbury

Abstract Australian rural women's health and well being is impacted by declining water availability to agricultural industries and rural communities. Rural women are subject to increasing demands on their time as a consequence of declining water and these demands are significantly impacting on women's health and well-being. This chapter reports findings from research undertaken in Australia's Murray-Darling Basin investigating the impacts of declining water on women's health and well-being. Impacts include being overloaded by work, caring and financial responsibilities and becoming increasingly financially responsible for family sustenance as farm incomes decline. There are increases in work related family separations and marital breakdowns and women report experiencing stress related symptoms such as depression. Findings indicate significant levels of family disruption and increasing incidences of violence against women associated with drought and income related stress. It is important that policy makers recognise these impacts so that specific and appropriate support services are provided in communities affected by declining water availability.

Keywords Climate change • Gender based violence • Women's health • Australia • Impacts

Background

Climate change has had a significant impact in Australia with an increasing frequency and severity of weather related events, some of which have had catastrophic consequences. In 2009, in Australia's worst recorded bushfire, 173 people died,

K. Whittenbury, BA, M.Litt., Ph.D. (✉)
Gender Leadership and Social Sustainability Research Unit, Monash University, Dandenong Road 900, 3145 Caulfield East, VIC, Australia
e-mail: Kerri.Whittenbury@monash.edu

M. Alston and K. Whittenbury (eds.), *Research, Action and Policy: Addressing the Gendered Impacts of Climate Change*, DOI 10.1007/978-94-007-5518-5_15,

5,000 were injured, 2,029 homes were destroyed, countless animals perished and over 4,500 km^2 of land was burnt (Australian Broadcasting Corporation 2009). The Australian state of Victoria experienced intense and widespread bush fires on a day that was so extreme in terms of heat and wind that a new classification system to account for such severe weather has since been developed (2009 Victorian Bushfire Royal Commission 2010). Within a month of Victoria's 'Black Saturday' bushfires, severe floods left a devastating impact on coastal Queensland. In late 2010 and 2011 Queensland was again inflicted with widespread devastating flooding caused by torrential rain. During the 2011 flood, 35 people died, nine remained missing and three quarters of the State was declared a disaster zone. At that time severe flooding also affected 50 communities in Victoria, some of which had been impacted by drought for a number of years prior. In early 2012 there was widespread flooding in New South Wales and Victoria with an area larger than the size of France inundated with water in New South Wales (Melbourne Age 2012). In addition to weather related disasters and catastrophic events, climate change can have more gradual, or incremental, effects and impacts. Incremental climate variability is exemplified by the widespread drought that occurred in much of Australia during the first decade of the twenty-first century (Alston 2011).

Australia's Murray-Darling Basin is known as Australia's food bowl, producing more than one third of Australia's food. Irrigated agriculture is a major source of production of food for Australia's domestic consumption and for the export market with 70% of Australia's irrigated agriculture located in the Basin (MDBA 2012). The research discussed in this chapter was conducted in 2009 and 2010 in Murray River communities in Victoria and New South Wales (see Basin maps). We anticipate undertaking further research in these Basin communities as Government policies that reduce water available to irrigated agriculture are implemented from 2012 (Figs. 15.1 and 15.2).

For the first decade of the twenty-first century a severe and widespread drought impacted on people, communities, industries and natural environments in Basin communities. Rainfall was extremely low and some rivers were reduced to a trickle or dried up altogether. Water supplies were severely curtailed and in many localities there were several years when there was zero water available for irrigated agriculture. The length and severity of the drought impacted on communities already experiencing the effects of major restructuring, which had been occurring for a number of decades (see Barr 2004; Hugo 2005; Lawrence 2005; Gray and Lawrence 2001; Pritchard and McManus 2000 for a discussion of rural industry restructuring and its impacts on rural communities).

Gender plays an important role in the way impacting factors are experienced and in the effects of those impacts. International research findings report that women are much more likely to die in disasters and are less likely to be accorded relief and reestablishment resources (Enarson 2012; Women's Environment and Development Organisation 2008). In Australian disasters such as bushfires and floods, women may be left to care for children, the elderly and livestock while men are absent fighting fires or maintaining flood defences (Parkinson 2011). Gender also affects experiences of slower onset change related to weather and climate such as drought

Fig. 15.1 Murray-Darling Basin, location in Australia (Source: Spatial Analysis Unit, Charles Sturt University)

(Alston et al. 2010) and impacts of structural changes such as industry restructures (see Alston 2006; 2005). Despite the evident differential gender experiences, we believe that these experiences are not adequately addressed in policy and service development aimed at addressing Australia's climate-based experiences.

Methodology

Building on earlier research (Alston and Mason 2008a, b; Alston and Witney-Soanes 2008; Whittenbury and Davidson 2009), three research projects have been undertaken in the Murray-Darling Basin since 2009 all focusing on the social impacts of drought and declining water. In 2009 we investigated the social impacts of declining water, on-going drought[1] and women's capacity and development needs

[1]Funded by Monash University

Fig. 15.2 Murray-Darling Basin – greater detail (Sourced from http://www.mdbc.gov.au/about/basin_statistics)

in relation to climate change[2] (see Alston et al. 2010). In 2010 we further explored the impacts for women raised by the 2009 research and undertook additional exploratory research investigating the impacts of declining water on women's health and wellbeing.[3]

A qualitative approach was used to capture people's experiences. For the 2009 studies 49 people (26 women and 23 men) in three predominantly irrigation communities close to the Murray River in Northern Victoria and Southern New South Wales that form part of the Murray-Darling Basin were interviewed. Initial interviews were conducted with key informants such as service providers working in support services, government agencies and authorities, health services, social

[2]Funded by Australian Government Department of Agriculture, Forestry and Fisheries
[3]Funded by Monash University

welfare agencies, education agencies and industry bodies. These interviews were followed by interviews with men and women involved in farming enterprises. In addition 24 women participated in three focus group interviews conducted in three communities chosen to reflect diversity in industry related to differences in agricultural commodities. The 2010 research was conducted in a regional city located in the Murray-Darling Basin and involved interviews with key informants about issues affecting women and women's access to, and use of, services. The findings discussed in this chapter draw on our 2009 and 2010 research.

Social Impacts

Our research into the social impacts of drought reveals significant hardship experienced by people and communities. Impacts include:

- reduced income and rising debt;
- a need for alternative income;
- a lack of work opportunities;
- health impacts;
- families forced to decide whether to stay in farming or leave;
- involuntary separation of families;
- stress, health and welfare issues;
- relationship conflict;
- community decline;
- loss of social capital; and
- depopulation – particularly of young people.

These impacts affect men and women in different ways and some factors impact more on men while others impact more on women. The focus of this chapter is on the impacts on women's health and well being of drought and declining water. Notwithstanding, it is important to note some of the key impacts on men remembering it is often women as partners, mothers, daughters and sisters who try to support men who are affected by drought and declining water. Impacts for men include:

- undertaking a majority of the physically demanding tasks, often undertaken by men working in isolation. This has potential health impacts for increased work related injuries and accidents as men attempt to undertake demanding tasks on their own. If injured, there is no one close by to render assistance or phone for an ambulance or emergency services;
- becoming more socially isolated;
- an articulated sense of 'emasculation', as they experience a loss of provider role;
- migrating away for work; and
- mental health and welfare issues including depression and suicide.

Our 2009 research found that social impacts for women include:

- an increased need to earn off-farm income;
- heavy 'other' workload on the farm, in the community, and in the household;
- monitoring the health of family and others;
- ignoring their own health needs;
- low levels of decision-making positions; and
- low acknowledgement of efforts.

We sought to further explore these impacts on women. In 2010 we undertook an exploratory study into the impacts of declining water on women's health and well-being in a regional city in the Murray-Darling Basin. For this research 11 key informants were interviewed, including industry representatives and service providers including health workers, welfare providers, financial counsellors, police representatives, youth workers and sexual assault workers. Indicating the complexity of rural work roles, seven of the eleven key informants reported personal involvement in farm enterprises; five were part of a farm family and two had sold their farms. Key informants were asked open ended questions about the impacts of drought and declining water, issues affecting women and about women's access to and use of services

Health and Wellbeing Impacts on Women

Findings from this study reinforce and elaborate on the impacts experienced by women that emerged during the 2009 research. Women's health and well being impacts include:

- being 'overloaded' with demands on time and resources;
- increasing work to earn income for necessities – including food;
- huge financial/lifestyle sacrifices to educate children;
- responsibility for men's (and family's) health and wellbeing;
- family disruption, separation for work; and
- increased violence against women.

Service providers and women participants reported that women are being 'overloaded' with demands placed on their time, efforts and resources. They are expected to undertake more tasks and assume more responsibilities leading to feelings of intense pressure and coping difficulties. The term 'at breaking point' was used by a number of respondents. One (female) manager of several women workers in a workplace setting reported that the women she manages are 'stretched to the limit'. If she asks them to do something extra such as work another shift to cover for staff absences, they are unable to do so and may become quite distressed because they are so overloaded. This can lead to stress related symptoms and feelings of being overwhelmed, which can result in depression and anxiety.

> I just cry at the drop of a hat (*Farm woman*).
>
> You would be surprised how many women are on anti depressants. **I'm not on them yet**. But there's close times. The only thing I do to, I (cannot) go to sleep sometimes, I just go to bed and I toss and I turn ... (*Farm woman*)
>
> I have stopped doing anything social. To do anything social is exhausting (*Farm woman*).
>
> When things are really bad they [women] don't want to come into town. Someone might ask them how they are and they might cry (*Service provider*).

In our research, we asked women what happens to their income. Their responses indicate that women's off-farm income is increasingly required to sustain living expenses and in many instances to cover business expenses previously paid for by the farm business such as electricity and telephone costs. In the past farm businesses earned the income to cover the necessities of family life and women's income was reportedly used for 'extras' such as family holidays. Now farm businesses are increasingly under financial pressure and unable to cover family sustenance so it is women's income that is sustaining families by providing the necessities of life such as food, clothing and utility costs. Women's off-farm income can thus be seen to financially 'prop up' farming as it enables otherwise financially unviable farming enterprises to continue to operate (see also Alston 2011). Women who previously may not have worked off-farm are now doing so and women who had hitherto worked part-time are increasing their paid work to full-time. Some of the men we interviewed in the 2009 studies suggested that their female partners work because they 'choose' to work but in exploring this further it became apparent that women's income was essential for these households.

> ...very important, extremely important (wife's off-farm income). Our kids wouldn't be getting the education they're getting if their Mum wasn't doing what she's done and been continuing to do. ...it's the driving force for both of us. I always knew (wife) would be working off the farm – that's who she is. ... but I didn't expect we'd be relying on it (*Farm man*).
>
> A lot of the men I've heard say over the years, "We wouldn't have survived on our farms without the extra income. It's made it much easier." So I think that's one of the most significant things (*Service provider*).

In addition to increasingly working off-farm, women's on-farm workload responsibilities are rising as farms shed labour in order to reduce input costs. Women reported that children are also more likely to be undertaking farm work. There is a growing tendency for women to arrange to take holiday leave from their paid (off-farm) employment during busy, or peak labour, periods on the farm so that they can undertake intensive farm work during their 'holidays'. Health professionals interviewed reported that this practice has potentially serious health ramifications for these women, who badly need a rest from the myriad of demands made on them.

Respondents also articulated that women are responsible for the health and well being of others, particularly family and partners. Women reported filtering information that they thought might exacerbate stress for their husbands including turning off electronic media when reports relating to drought and declining water were aired. Some of the women interviewed were very concerned for their partners' safety and psychological health because men are increasingly working in isolation as they have cut back on the cost of employed labour. Women raised concerns about their partner's heightened risk of accidents and injury and lack of available assistance due to men working on their own.

> So if one of your family members are depressed, well then you're the supporting person more often and you have to rally into the family again instead of getting on with your own life and different things like that. And I've spoken to a few of the men, and they are indecisive (*Farm woman*).

Women may also be responsible for elder care, including that of their partners' parents who may live nearby. Traditional values regarding gender roles, which are common in rural communities, exert strong expectations on women to bear the responsibility of family care giving and to carry out the tasks involved in such care giving (Whittenbury 2004).

> And look after these parents as well, so they've not only got their kids they've got ageing parents-in-law who are stubborn as buggery who might live out on the farm who, "No, we're not coming into town". So the daughter-in-law, there's a big expectation for her to then look after his parents, huge… (*Service provider*).

Health care workers and other key informants reported that the heavy demands and increasing workloads being placed on women are causing women to neglect their own health care needs and are having adverse health consequences in other ways. In our field research we spoke to women well past retirement age, who continue to work because of financial necessity, often in physically demanding roles such as nursing and aged care work in addition to continuing to undertake farm work. These women reported physical symptoms such as back and joint pain and other symptoms associated with extreme fatigue. It is not only older women who are impacted in this way but younger women, who need to increase their paid off-farm work and unpaid farm work.

> She pays the grocery bill so the kids are fed, everyone's fed, everyone's happy …but if you look at her schedule – they're stand up shifts … she became a wreck and now she's had a few little ailments and she's in a fair bit of pain… (*Farm man*).

What has also become clear in our research are the huge financial and lifestyle sacrifices made by women to educate their children. Again and again respondents spoke of the overwhelming efforts women are making to educate children – particularly for tertiary education, which usually requires the children to live away from home with parents funding much of the associated costs, which can be up to $20,000 per year per child if the children are at a fee charging education provider. Informants report women working extra time, cutting back on their own personal expenditure and 'going without' in order to ensure their children are educated so that they have alternative options available to them rather than farming (see also Alston et al. 2010).

> I've always been the off-farm income …plus it keeps me sane. Because I think if I had to sit out here and watch the drought I'd go stark raving mad. … there was an undercurrent of

'poor Husband'. ... but who else is going to pay for the kids' education? ... [my income] takes the pressure off school fees, private health insurance – all the extras I suppose. ... My income gives us a bit of – well pride! (*Farm woman*)

But predominantly if you talk to most wine grape growers, they don't want their son to take over the block anyway, or daughter. And most times the children don't want to take over (*Farm woman*).

Family Disruption and Separation

Respondents report an increase in family separation for significant periods as people go away from home in order to obtain paid employment. The mining and minerals sector plays a major role in this type of employment with employees work patterns involving fly in, fly out arrangements. The most common scenario is that one spouse, usually the husband, goes away for work leaving the woman at home to care for children, and perhaps elders, maintain the home and continue to run the farm in addition to working off-farm.

Contracting is another type of employment that involves lengthy periods of working away from home. Those involved in contracting – mostly men, may have their own 'family farms', which they leave sometimes for several months at a time to undertake seasonal tasks such as sowing, planting and harvesting, often using their own machinery and equipment, which is commonly under finance. Again, in this scenario, women are left with a significant workload and are responsible for children, home and farm as well as working for income. This contributes to women's overload, with consequent health and wellbeing repercussions.

We also learned of instances where it is the female spouse who works away. In some cases, women worked long distances from the home and returned each weekend. This involves many hours of driving, often at night, on poor quality rural roads to be met with an expectation of undertaking farm-work on the weekends. In other instances, women and children rented a house in nearby towns or provincial cities so that women could work and children could attend school and extra-curricular activities without the need for long commutes and expensive fuel costs. Such separations place considerable stress on the marital/partner relationship with several respondents using the term 'living parallel lives' to refer to their relationship with their partners.

[My wife] is a lot ... [angrier] than she used to be and I see a grumpy person now instead of the person who was the fun-loving housewife that she used to be. Because she's raising kids obviously she can't have a full-time nine to five job ... it's not a normal life, you're living an alternative life (*Farm man*).

If you want your marriage to continue then you have to start looking after yourself and almost live a parallel life for a while so that you're able to cope. He doesn't get it. He's just stuck up there (on the farm) by himself and its all falling down around him like a deck of cards – that's how I see it. I really have a completely separate life down here (city) and without that I'd be in a funny farm (*Farm woman*).

> [I moved to the city for education] because we couldn't afford to send (children) away (to boarding school).... I was there for four years. (Husband) would get down when he could. We would get up [to the farm] mainly in school holidays ... it wasn't the ideal situation but it was better than boarding school. I did that for four years working part-time as a nurse (*Farm woman*).

Key informants who work with these families suggest that couples have a perception that the income earned from working away will ameliorate problems caused by lack of income and increasing debt associated with drought and declining water. Service providers suggest that this practice is not sustainable in the longer term as it is disruptive for families and communities and can lead to more problems, particularly marital difficulties that may culminate in separation and divorce.

> Men mainly [working away], which is causing – if the female presents [to the service provider], which is generally the case, then that will be because they're struggling with "he's away and I have to do all this", and then he's back and they can't quite get it to balance again, so different rules for when they're here and there, and the stress of them being away, and ... then they come home and the female might want a break but it's kind of – the male sees it as his break coming home. It's just an assumption that they'll go away and work, they'll bring back lots of money and then we'll be better off, but people don't think about what it means to have a family separated for periods of time, blocks of time. So mum's mum and dad when dad's away, and then how does dad – what's his role when he gets back? And sometimes I think that's a big issue for the male to know where they fit back into the family where they feel a bit like the third wheel (*Service provider*).

Migration is not new in a global context and has often occurred as a result of economic need but it has not historically occurred in Australia, particularly in the context discussed above. Australian farm families have previously been financially secure whereas now, increasing numbers of farm families are in precarious financial circumstances, which limits their available options (Dibden and Cocklin 2005; Pritchard and McMannus 2000).

Violence Against Women

Researchers investigating gender and disasters have found that violence against women increases following disasters (Enarson 2012; Parkinson 2011; AusAID 2008). While this has been mostly reported in the context of developing countries, there is emerging Australian and international evidence that in developed countries violence against women increases following disasters and severe weather events (David and Enarson 2012; Enarson 2012; Parkinson 2011). Queensland service providers reported an increase in requests for assistance from women who had experienced partner violence following the 2010 and 2011 floods (ABC 2011) and in Victoria, women's service providers reported increased violence against women in the aftermath of the 2009 bushfires (Women's Health in the Goulburn North East 2012). In New Zealand, a spike in domestic violence reports occurred following the 2010 Canterbury earthquake and again 6 months later following the 2011 Christchurch earthquake (Parkinson 2011). Anastario et al. (2009) report 'elevated rates' of violence against women and sexual assault following the Hurricane Katrina disaster, which affected New Orleans in 2005 and that these 'elevated rates' continued for the 2 years post-Katrina for which they collected data (p22).

International development agencies are well aware of instances of increased violence against women following disasters and seek quantitative 'hard' evidence in order to influence policy development in this area (Oxfam Australia, 2011, personal communication; UNWomen, 2011, personal communication; WeDo 2011). In a recent literature review of gender, disaster and violence, Parkinson (2011) highlights the under-reporting of violence against women in disasters in both developing country and developed country settings.

While there is some research being undertaken into violence against women following disaster, what is less documented is violence against women associated with incremental or gradual climate change and climate variability. Our research in the Murray-Darling Basin found emerging evidence that gender based violence increased during severe drought years. Service providers believed that tension exacerbated by financial pressures associated with drought led to gender based violence. Service providers also reported that men increased drug and alcohol consumption as a coping device and that this led to increased violence. While gender based violence encompasses a variety of behaviours, service providers mentioned instances of emotional abuse, financial control and abuse and isolating women. Isolation was often associated with financial control as women were not allowed to drive to town to shop, socialise or visit friends because their partners' cited the high cost of fuel. In such instances male partners would buy groceries and do other necessary tasks when they went to town on 'legitimate' business.

> The financial difficulty. You'd say, "What's the issue here?" "Financial difficulties. We're just struggling." And what I mean by people that you wouldn't usually see isn't necessarily that it wouldn't be someone on the farm. We might have a repeat address that we go to that is a farm, but it would be people that we have never had any reports of family violence before... (*Service provider*).

> I just focus on family violence. Going back probably from the end of 2008, start of 2009, we were really hit by the drought and things were quite bad. I did notice an increase in family violence – not family violence physical assault, more your verbal altercations between families that you probably normally wouldn't see or have any reports of family violence... (*Service provider*).

> I think one of the most significant, we all know that drugs and alcohol can be contributing factors to domestic violence, but I would say financial pressure is one of the most significant factors. And we see that when the drought was on, people walking off properties, violence was escalating and there were a number of suicides and particularly the males were suiciding. But lots of depression amongst the women on a bigger scale. I would say anecdotally that we saw increases (*Service provider*).

> And the blokes say no, well we can't afford the petrol and I need to go and see so and so, so I'll get the shopping while I'm in there [town]. So that's a form of domestic violence (*Service provider*).

The service providers interviewed reported that instances of violence and intimidation were rarely raised by women directly. Instead service providers needed to remain alert to indications that violence and abuse might be an issue. Sometimes when working with couples, they observed interaction between partners that was indicative of intimidation and violence. Service providers noted that they would then try to arrange another appointment when the woman was on her own so that they could work with the woman or refer her to specialist services.

> Most of the time you just observe it [intimidation]. It's very rarely raised. And it probably takes a lot of relationship building before anyone would raise that kind of thing. It's usually something that you observe, and I normally try and refer those off in particular . . . (*Service provider*).

> You find a lot of that . . . it's not always that easy to see – mental [abuse]. I go to farms quite a lot, and the man will do all the talking, and the wife won't say anything. And a lot of the time I pick up from that she's not game to say anything basically. When that happens, I usually try to make another appointment to make sure he's not there. And you usually get a lot more, and you do, you start to pick up that – and quite often it's things like, because she might want to say, "Well look, I just think we're going down the gurgler," but he doesn't want her to say that to anybody. Those sorts of things. And yeah, I see that quite a lot. Myself, I would probably say more the mental abuse. (*Service provider*)

Women who experience gender based violence may experience other symptoms and seek professional help for these symptoms. Health professionals and other service providers report that women may initially present with depression or anxiety – and that further probing by the service provider identifies that gender based violence is an issue. Service providers further report that they need to be conscious of the language and terminology used when working with these women as they do not identify with the current terminology, such as 'abuse' and 'violence', and that women do not respond to this terminology, which may cause them to retreat from potential help. According to service providers, women who experience abuse may not recognize it as abuse and may be reluctant to report abuse as a crime to police.

> Very few will go to the police in the small communities (*Service provider*).

> For women that are on particularly farming, the dry land farming and certainly out on isolated areas, to actually present and even get away sometimes is a major problem, and even to come an access information. But mostly they're struggling on. They might present to GPs (local doctors) over again with depression, but the questions they've asked as to what's really going on, it's obviously often family violence and I would say if there is a tendency for that anyway, the financial pressure, having to walk off or be in huge amounts of debt, that puts an enormous amount of pressure on people. That is a regular thing (*Service provider*).

> To use the terminology, "I think you're being abused and I think that you're in a domestic violence situation," is really even worse for them than – they'll just move it away again (*Service provider*).

> Absolutely. As soon as the [term] domestic violence was used it was you could actually physically see the door shutting, and the conversation changed . . . (*Service provider*).

Conclusion

The impacts of drought and declining water on women's health and well-being cannot be considered in isolation from other complex intervening factors such as rural restructuring, falling commodity prices and market deregulation. In addition, considerable permanent reductions to the availability of irrigation water are being implemented across the Murray-Darling Basin (see Alston and Whittenbury 2012; forthcoming). These reductions will affect some communities more than others and the Murray-Darling Basin Authority in its Draft Plan for the Basin has identified five regions as being 'especially vulnerable' to changes in water availability that will occur as the Plan is implemented (MDBA 2011 p 123). People in these communities will be especially vulnerable to the impacts outlined above and it can be expected that women will be particularly vulnerable due to the hardship and financial stress that will occur as a result of the Plan's implementation. It is important that service providers and policy makers are aware of the social, health and well being impacts identified so that appropriate supports and services are established in the affected localities. The reports of increases in gender based violence occurring in relation to economic and social stress associated with incremental climate change are extremely concerning. Adequate and relevant services to address gender based violence need to be targeted to women in affected communities.

The long standing severe drought that affected the localities that were the subject of our 2009 and 2010 research ended in late 2010 yet the impacts experienced by people and communities in these areas continue. Many of these areas have been impacted by severe flooding in the 2 years since the drought ended resulting in crop and livestock losses and crop downgrades. Australia's high value dollar, compared to a number of international currencies, has impacted on farm exports, which in turn impacts on the incomes of those farm enterprises still servicing debts incurred when they were in the grip of severe drought. Major structural changes such as depopulation and local business closures that occurred during the drought do not simply reverse because the drought has ended. Our investigations indicate that the after effects of such a prolonged drought have impacted people's and communities' resilience resulting in 'community post-traumatic depression' as one key informant described it. It cannot be presumed that because the drought has broken, people and communities are no longer impacted by its effects. Policy makers and service providers need to be aware of the continuation of these impacts and to implement and continue supports, services and other interventions to assist people and communities adapt to the changes that have occurred and will continue to occur.

References

2009 Victorian Bushfire Royal Commission, Final report, Recommendations July 2010 http://www.royalcommission.vic.gov.au/Assets/VBRC-Final-Report-Recommendations.pdf. Accessed 12 Jan 2012

Alston M (2005) Social exclusion in rural areas. In: Cocklin C, Dibden J (eds) Sustainability and change in rural Australia. UNSW Press, Kensington
Alston M (2006) The impact of drought on gender and power differentials in Australian farm families. In: Bock B, Shortall S (eds) Rural gender relations: issues and case studies. CABI Publishing, Oxfordshire
Alston M (2011) Gender and climate change in Australia. J Sociol 47:53–70 http://jos.sagepub.com/content/47/1/53
Alston M, Mason R (2008a) Who determines access to Australia's water? Gender, citizenship and stakeholder priorities in the Australian water crisis. Rural Soc (Special edition Gend Water) 18(3):214–219
Alston M, Mason R (2008b) Who turns the taps off? Introducing social flow to the Australian water debate. Rural Soc 18(2):131–138
Alston M, Whittenbury K (2012) Climate change and water policy in Australia's irrigation areas: a lost opportunity for a partnership model of governance. Environ Polit 20(6):899–917, Routlege
Alston M, Witney-Soanes K (2008) Social impacts of drought and declining water availability in the Murray Darling Basin. Institute of Land, Water and Society, Charles Sturt University, Wagga Wagga
Alston M, Whittenbury K, Haynes A (2010) The social impacts of declining water availability and ongoing drought in the Murray-Darling Basin: short report. Gender, Leadership and Social Sustainiability Research Unit, Monash University, November
Alston M, Whittenbury K Social impacts of reduced water availability in Australia's Murray Darling Basin: adaptation or maladaptation. Int J Water (forthcoming)
Anastario M, Shehab N, Lawry L (2009) Increased gender-based violence among women internally displaced in Mississippi two years post-Hurrican Katrina. Disaster Med Public Health Prep 3(1):18–26
AusAID Office of Development Effectiveness (2008) Violence against women in Melanesia and East Timor building on global and regional promising approaches. Office of Development Effectiveness, Canberra. Cited in Parkinson D (2011) Gender, disaster and violence; literature review. Women's Health Goulburn North East, Victoria
Australian Broadcasting Corporation – ABC (2009) Online http://www.abc.net.au/innovation/blacksaturday/#/stories/mosaic
Australian Broadcasting Corporation – ABC (2011) Online http://www.abc.net.au/radionational/programs/breakfast/floods-make-life-tougher-for-qld-domestic-violence/2998262
Barr N (2004) The micro-dynamics of occupational and demographic change in Australian agriculture: 1976–2001. Australian Bureau of Statistics, Canberra
David E, Enarson E (eds) (2012) The women of Katrina: how gender, race and class does matter in an American disaster. Vanderbilt University Press, Nashville
Dibden J, Cocklin C (2005) Introduction. In: Cocklin C, Dibden J (eds) Sustainability and change in rural Australia. UNSW Press, Sydney, pp 1–18
Enarson E (2012) Women confronting natural disaster: from vulnerability to resistance. Lynne Reinner Publishers, Boulder
Gray I, Lawrence G (2001) A future for regional Australia: escaping global misfortune. Cambridge University Press, Cambridge
Hugo G (2005) The state of rural populations. In: Cocklin C, Dibden J (eds) Sustainability and change in rural Australia. UNSW Press, Sydney, pp 56–79
Lawrence G (2005) Globalisation, agricultural production systems and rural restructuring. In: Cocklin C, Dibden J (eds) Sustainability and change in rural Australia. UNSW Press, Sydney, pp 104–120
Melbourne Age (2012, 2 March) http://www.theage.com.au/victoria/defiant-locals-fight-to-save-homes-from-floods-20120301-1u615.html
Murray Darling Basin Authority website. http://www.mdba.gov.au/. Accessed 1 Mar 2012
Murray-Darling Basin Authority (MDBA) (2011) Draft basin plan. http://www.mdba.gov.au/draft-basin-plan

Parkinson D (2011) Gender disaster and violence; literature review. Women's Health Goulburn North East, Victoria
Pritchard B, McManus P (eds) (2000) Land of discontent: the dynamics of change in rural and regional Australia. UNSW Press, Sydney
Whittenbury K (2004) Exploring the taken-for-granted: men's and women's perceptions of family life in a rural community. PhD thesis, Charles Sturt University, Wagga Wagga
Whittenbury K, Davidson P (2009) Beyond adoption: the need for a broad understanding of factors that influence irrigators' decision-making. Rural Soc 19(1):4–16
Women's Environment and Development Organisation (2008) Gender, climate-change and human security: lessons from Bangladesh, Ghana and Senegal. WeDo, New York
Women's Environment and Development Organisation [WeDo] (2011) Personal communication
Women's Health in the Goulburn North East (2012, March) http://www.whin.org.au/images/PDFs/the_way_he_tells_it.pdf

Chapter 16
Women Farmer Scientists in Participatory Action Research Processes for Adaptation

Bettina Koelle

Abstract Increasingly there is international pressure to engage in adaptation to address, or preferably to prevent, negative impacts of climate change. And although some downscaled forecasts are available, they often fail to link the local knowledge and livelihood systems crucial for successful adaptation. In the Suid Bokkeveld men and women farmers and scientists embarked on a scientific research partnership: jointly developing research questions, monitoring and discussing results, which were then reported back to members of the community at quarterly community workshops. This chapter explores the mechanisms that supported collaboration between rural women and scientists and makes recommendations on how action research can support women infarmer science and adaptation.

Keywords Adaptation • Climate change • Action research • Farmer science • Women farmers

---

Bettina Koelle graduated with a Masters degree at the Free University of Berlin (Germany) in 1997 and has been working as a development practitioner in the field for more than 15 years. With a strong passion for participatory approaches she has supported the development of various participatory methodologies to support an action research approach linking grassroots processes to global debate, specifically focusing on linking local adaptation action to the climate negotiations and the IPCC process by providing input and case studies from the field. In this context her focus has been on marginalized groups and especially women.

She is currently undertaking research towards her Ph.D. at the University of Cape Town (South Africa) on participatory monitoring and evaluation of community based adaptation projects.

B. Koelle, M.Sc. (✉)
Department of Geography, Free University of Berlin, Germany

Department of Environmental and Geographical Science, University of Cape Town, Cape Town, South Africa

Indigo Development & Change, Nieuwoudtville, South Africa
e-mail: bettina@indigo-dc.org

M. Alston and K. Whittenbury (eds.), *Research, Action and Policy: Addressing the Gendered Impacts of Climate Change*, DOI 10.1007/978-94-007-5518-5_16,

Introduction

In the South African context, negative effects of climate change on agriculture are already observed and the predictions state that the impacts will be more severe in the future (Engelbrecht and Landman 2010; Lotter 2010). Although some uncertainty persists regarding appropriate adaptation options in South African agriculture, it is anticipated that altered climatic conditions will pose more challenges for South African agriculture (Lotter 2010).

While there is certainly pressure to adapt to future climate scenarios, climate variability and change is already affecting men and women in South Africa, placing additional stress on their often already vulnerable situations. Due to the uneven distribution of roles and responsibilities between men and women and the often restricted access to assets and resources by women gender differentiated impacts of climate variability and change are already observable (Babugura 2010).

In order to understand the gender dynamics in South Africa it is important to consider the historical context: South African society has made the shift from the oppressive Apartheid regime towards democracy with the first elections in 1994 – and although progress has been made, strong injustices persist to date. This relates especially to the distribution of resources (e.g. arable farm land) and the social consequences of oppression. Climate change poses an additional challenge – and is likely to have a negative compounding effect on the livelihoods of already marginalized groups. This case study explores some of the patterns and mechanisms to promote adaptation that is rooted in participatory processes and examines the extent to which a Participatory Action Research process can support women "grassroot-scientists" on-farm in their personal transformation processes. The aim is to define a methodology that not only promotes transformation but also supports anticipatory adaptation by vulnerable groups. This is supported by the definition of the IPCC, which emphasises that the adaptation of human systems should be based on a process of adjustment to actual or expected climate in order to either moderate harm and/or to exploit beneficial opportunities (IPCC 2012).

From Adaptation Towards Transformation

Adaptation has been framed in various ways – and increasingly resilience and the protection of vulnerable groups have been emphasised (Schipper 2007; Schipper and Burton 2009; Vogel et al. 2007). While the term resilience is often used to describe the adaptation pathway towards sustainable livelihoods (Schipper and Burton 2009), Pelling argues that an adaptation typology should distinguish between transformation for resilience (maintaining the status quo), transformation for transition (with some positive changes to the status quo) and adaptation for transformation (radical change towards sustainable livelihoods) (Pelling 2011). Adaptation for transformation also addresses important justice issues, as often it is not desirable for the actors to either regain or maintain the former or current status quo.

As international funding for implementing adaptation measures is increasingly leveraged, it is crucial to consider the best pathways towards transformative adaptation – and to ensure that marginalised groups are supported in their self-determined development. In order to understand the stressors and vulnerabilities, but also the visions and ambitions of vulnerable groups, a range of tools for gender analysis have been developed, focusing on three areas of investigation: gender specific on and off farm operations and activities, understanding control over resources and production (supported by gender disaggregated data) and lastly the implications of the livelihood strategy for transformation and innovation (Feldstein and Jiggins 1994).

Using these or other participatory methods for gender analysis, it is crucial to learn from men and women alike, to ensure that systems can be explored in their complexity. Many biases come in to play when collaborating with stakeholders: amongst others gender dynamics, middlemen and gate keepers – which might prevent broad participation and exclude marginalised groups (Sutherland 1994). The selection of the location of research sites for the farmer collaboration is often subjected to the urban bias: proximity to urban areas and good infrastructure allowing easy access will lead to a skewed bias towards potentially better resourced and connected communities (Chambers 1983; Sutherland 1994). This has been widely discussed in the development discourse, but must be considered for the adaptation debate.

The Participatory Action Research process should address the questions of local farmers – and participatory processes can encourage broad participation, including of women and marginalised groups within the community (Buenavista and Flora 1994). Vogel et al. (2007) argue that collaboration between local knowledge bearers and scientists should become a matter of priority to address the adaptation challenge (Vogel et al. 2007). This co-produced knowledge is important because it is jointly developed, tested and improved, and might just bear the key to address the adaptation challenges of the future.

Participation has been supported as an effective tool for transformation in society: *"Its Philosophy is founded on principles of peace, justice and equality, a profound belief in the worth of everyone and the sanctity of the natural world"* (Ledwith and Springett 2010).

In the following section the possible mechanisms to support such learning processes are explored – bearing in mind that the aim of the adaptation process is not just a technical but more profoundly a personal transformation process towards more sustainable livelihoods.

Case Study: Women Farmer Scientists in the Suid Bokkeveld

The Study Area

South Africa is a country already detecting change in climate and extreme weather events (Engelbrecht and Landman 2010). However the Western part of the country has a long history of experiencing extreme climate variability: since the earliest written records severe droughts have been recorded in the area. However, these extreme spells seem to have increased in frequency in the past decades (Archer 2006).

The case study area, Suid Bokkeveld, is situated in the South-Western corner of the Northern Cape Province – the largest South African Province (29.7% of South Africa's land surface) with the lowest population density in the country of just 2.27 persons per km^2 (Statistics South Africa 2003). The name *Suid Bokkeveld* is an Afrikaans phrase (first language to almost all inhabitants in the Suid Bokkeveld today), and literally translates as "Southern Antelope Veldt" referring to the vast herds of indigenous antelope that used to roam the plateau after good rains. Situated in a winter rainfall area, the wet and cold winter months bring most of the rainfall in a normal year; 350 mm per annum for Nieuwoudtville. However in drought years, the rains can fail – sometimes for successive years – affecting farming families in the area severely. The annual average rainfall varies strongly from year to year along a North–south gradient on the Plateau (Fig. 16.1).

The vegetation in the area includes two biomes identified as global biodiversity hotspots, namely the Succulent Karroo on shale and the Cape Floral Region (CFR) on Table Mountain Sandstone soils. The agricultural livelihood strategy of small scale farmers in the area consists of some general farming activities (such as keeping chicken and pigs, cropping, vegetable farming for own consumption) but the main sources of cash income are derived from farming with livestock and rooibos tea (a herbal infusion made from a plant endemic to the CFR: *Aspalathus linearis* and exported worldwide).

The Suid Bokkeveld consists of agriculturally marginal land of extremely low nutritional status and while the farmers rely on fountains and boreholes to provide water for household use and for providing water for livestock, erratic rainfall over some years led to a severe drought between 2003 and 2006, and a crisis in water supply on some farms in the Suid Bokkeveld (Own interviews).

The inhabitants of the Suid Bokkeveld live in scattered smaller mostly family settlements in proximity of a permanent water source. The distance between settlements is, on average, more than 50 km on poorly maintained dirt roads. Transport is a key challenge as very few inhabitants own their own transport and public transport is absent from the area. Nieuwoudtville is the centre for business and supplies and also houses the offices of two NGOs and the Heiveld Co-operative, a local small-scale farmer organisation.

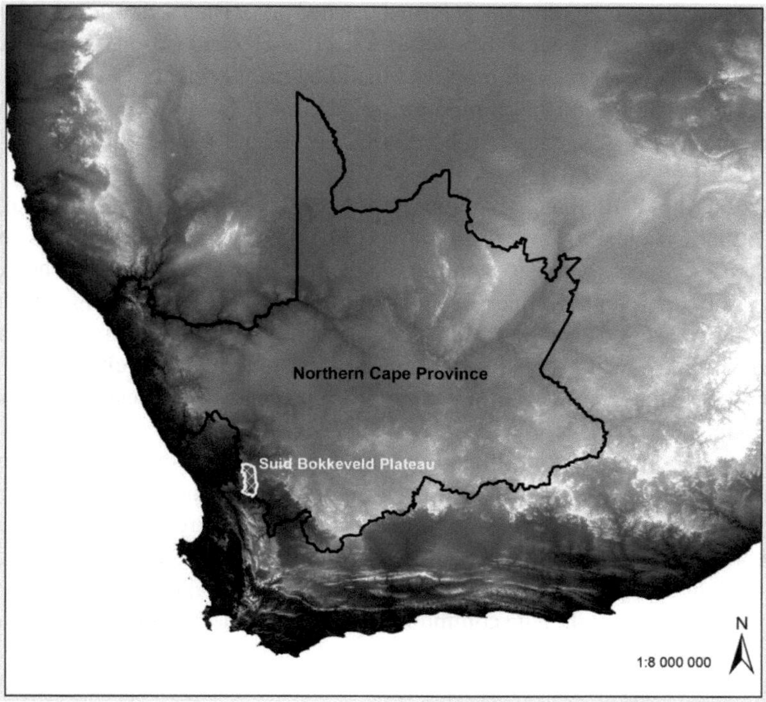

Fig. 16.1 Northern Cape Province (South Africa) and the Bokkeveld Plateau on the Western Edge of the Great Karoo Plateau (Map: B Koelle, using data from South African Surveyor General)

The Participatory Action Research Process

The research approach described in this chapter is firmly rooted in Participatory Action Research (PAR), aiming at transformation and supporting marginalised groups (Kemmis and McTaggart 2000; Senge and Scharmer 2001). The on-going Participatory Action Research process with the local Suid Bokkeveld community, facilitated by two South African NGOs (Environmental Monitoring Group and Indigo development & change), has been documented since the first interaction in 1999 (Archer et al. 2008; Koelle and Oettlé 2009; Oettlé 2006; Oettlé et al. 2004; Oettlé et al. 2009; Oettlé and Koelle 2003; Oettlé 2005)

The farmers in the Suid Bokkeveld with support from the above NGOs decided to make a new beginning after the end of the Apartheid regime: they embarked on a Participatory Action Research process in a participatory workshop in 1999, focusing on a number of challenges identified ranging from poor access to markets for their main cash crop, rooibos tea, to high unemployment rates within the community and incidents of domestic and gender-based violence. The participatory process led to a range of activities. Founding a new farmer–owned organisation that accessed the fair trade and organic markets for their rooibos tea (the Heiveld Co-operative Ltd.) was a pivotal move. Other parallel processes included management of natural resources and climate change adaptation monitoring processes, farmer experimentation and collaboration with scientists (Archer 2006; Koelle and Oettlé 2009; Oettlé et al. 2004; Oettlé et al. 2009).

Initially the Participatory Action Research process was focused on rooibos production and creating improved market access for small-scale rooibos farmers in the area. Farmers wanted to increase production, improve their market access and realise stable and higher prices for their rooibos tea. The founding of the Heiveld Co-operative in 2000 enabled the small scale farmers to access fair trade and organic markets directly, while creating an organization that was entirely owned by its members and thus was also in a position to represent the interests of small scale farmers in general.

Of special concern to the founding members of the Co-operative was gender balance. Extensive discussions took place from the beginning to find ways to ensure broad participation of women in all processes. Policies were put in place to ensure that women would also become members of the Heiveld Co-operative in their own right, and that active capacity development of men and women would ensure sound social development within the community and sustainable organic farming practices alike. This was a significant step in the development of the organisation and it shifted gender dynamics within the Co-operative and in the Suid Bokkeveld community. These formal policies and processes were important to ensure that women were not just feeling welcome to these processes, but felt that their unique contributions towards the development processes were highly sought after and valued.

Increasing Women's Participation by Listening and Addressing Special Needs

Climate Change Preparedness Workshops were initiated to share results and discuss possible adaptation strategies. More specifically the workshops were focused on the following elements: sharing adaptation stories, both local and from other places (Southern Africa and globally), discussing and sharing action research ideas and findings, sharing climate monitoring data from different farms and comparing these, compiling seasonal climate calendars and discussing the seasonal forecast for the next 3 months and possible actions to be taken in response.

Analysis of past workshop documentation suggests that a key turning point was reached when Indigo introduced a parallel workshop for children that allowed women to fully participate while their children were engaged in the parallel workshop. The format of the workshops was changed in such a way that the beginning of the workshop was held with everybody together, whereafter the children proceed to their own workshop. Towards the end of the morning the children gave a short report back about their activities to the larger workshop. Each workshop ended with a reflection of all participants to evaluate the workshop day – and significantly many women have stated that what they liked most was the fact that the children had the courage to present their learnings and projects in front of the entire group. The informal and friendly atmosphere of the workshop was appreciated, as was the interactive format, including stories from other countries, experiments conducted and group work to reflect on past climate. The facilitation team created the space for this reflection and the opportunities it provided to understand the expressed needs of the communities. The children's workshop was suggested by the children themselves, they expressed an interest in "doing something too". Budget constraints and inflexible processes and facilitators can often mean that we miss out on opportunities "stumbled upon" like this one.

Women Farmer Researchers in the PAR Process

The research questions and the participation in these research processes changed over time: while in the beginning mainly male farmers participated in the on-farm experiments and collaborated closely with scientists, over time more and more women participated in the quarterly climate change preparedness workshops and engaged in research.

While the Participatory Action Research process and the organisational development of the Heiveld Co-operative were closely linked and feeding into each other, a series of PAR processes implemented by farmers can be singled out to demonstrate the development of the research activities over time (Table 16.1). The research design, implementation and findings were shared and discussed in the quarterly workshops and led to the development of new research ideas and adjusted farming practices. Increasingly women participated in these processes, and some processes were entirely led by women. The term 'women farmer researchers' has been chosen to illustrate the significant contribution women make while contributing their knowledge and creativity to research processes. It also emphasises that research that encompassing local knowledge can be scientifically rigorous and that it is possible to integrate local and scientific knowledge. The term 'researcher' also seeks to clarify that we do not see women farmers as mere helpers or field assistants in the research process but as equal partners driving the process (Fig. 16.2).

The action research projects undertaken and the partners involved are outlined in Table 16.1.

Over time, more and more women participated. This shifted slowly at first, but dramatically increased momentum. After some women did well with the research and reported back to the community, the participation of women increased dramatically.

This example shows that often we need to listen closely to understand gender dynamics and make adjustments in process and methodology that allows us to create the space and environment necessary for effective participation by marginalised groups (Fig. 16.3).

Table 16.1 Agricultural action research processes in the Suid Bokkeveld since 2000 and the partners involved: University of Cape Town (UCT), University of Stellenbosch (USB), Environmental Monitoring Group (EMG), Indigo Development & Change (Indigo), Heiveld Co-operative Ltd. (Heiveld), Council for Scientific and Industrial Research (CSIR) Council for Geoscience South Africa (CGS)

Year started	Research topic	Farmer researchers	Collaborating researchers	Status
2003	Sustainable harvesting of rooibos (wild and cultivated)	Male farmers	UCT EMG	Concluded
2006	Germination of wild rooibos	Male farmers and one woman farmer	UCT USB EMG Indigo	Ongoing
2008	Water monitoring on farm	Women farmers (and some male farmers)	UCT Indigo CGS	Ongoing, special focus on women farmers
2008	Erosion control and organic farming	Men and women farmers	Heiveld EMG	Ongoing
2010	Climate monitoring and climate diaries	Two men and three women farmers	Indigo UCT	Ongoing
2012	Heat stress and livestock: thresholds	Women farmers	CSIR UCT EMG Indigo	Planned

Owning the Data Collected: Climate Diaries

The introduction of climate diaries in 2010 by Indigo Development & Change has further complemented the long-term weather monitoring through automated weather stations on four different farms in the Suid Bokkeveld. Weekly recording of maximum and minimum temperatures, observations and current and planned farming activities are recorded on a weekly basis by so called climate monitors and then compared to the seasonal climate forecast. The farmers who have committed themselves to this process were not paid to maintain the climate diaries. When asked by an external evaluator if they were financially compensated, one farmer replied: "*Why should we be compensated for something that benefits us*?" (Interview with women farmer, own records). This illustrates that the farmers have ownership of the learning process and do not feel they need to be financially compensated for their involvement in PAR processes. The method of using climate diaries allows the farmers to be in charge of the data they are collecting: the farmers remain in possession of their diaries and Indigo staff photographed the diaries every 3 months for a long term record (Fig. 16.4).

Fig. 16.2 Drieka Kotze on the Farm Melkkraal was one of the first women farmer scientists and is now the "godmother" of an automated weather station and an active PAR researcher (Photo: B Koelle)

Fig. 16.3 Katriena Fortuin on the Farm Melkkraal and her son: enabling women to participate by "stumbling upon" the solution (Photo: B Koelle)

Fig. 16.4 Karien Kotze with her climate diary on the Farm Melkkraal: owning process and data (Photo: B Koelle)

Women Initiating Research: Water Monitoring

Water was identified, especially by the women, as a key resource to monitor in order to anticipate change in water supply and to be able to take appropriate action. The PAR process included participatory methods such as Participatory GIS (PGIS) and Participatory Video (PV). Both methods have been used in the development context, and create ways for men and women to express their differing views and to support an active learning process (R Chambers 2005; Lunch 2007). The use of technology was quickly taken up by the male farmers and once smaller workshops were introduced, many women engaged with the technology, handling video cameras and GPS devices with confidence. The water monitoring is currently undertaken by six women on different farms in collaboration with Indigo and the Council for Geoscience.

Women in PAR Processes for Adaptation

The example of the Suid Bokkeveld has shown that participatory action research processes can offer opportunities for women to effectively engage in adaptation processes. These processes have to be culturally sensitive and take into account the very specific conditions of each community, especially power relationships and access to assets, visions and ambitions and the natural resource base.

In order to more effectively engage women as action researchers in adaptation, it is important to clearly understand the challenges men and women encounter. A list of challenges, aims and possible processes or tools to illustrate possible pathways for implementation has been compiled (Table 16.2). In order to illustrate possible pathways for transformation, to address injustices and support marginalised groups it is crucial to gain a profound understanding of the layered complexity of the situation and to ensure that the intended beneficiaries are in fact the actors and drivers of this process. In the described case study the quarterly climate change preparedness workshops have been an important place to share results from ongoing research processes and the workshops have provided a platform for discussing new research ideas. Once the research questions had been formulated, farmers and scientists undertook the research together. This is a delicate process and requires careful gentle facilitation, that provides farmers and scientists alike the space to contribute in an environment where they feel safe.

While designing the research it is important to explore and understand the research environment – and the different roles men and women will have in this particular community. Ongoing participatory action research processes often require continuous recording of research data, and because women are often less mobile while looking after family members at home, they are often strong research partners.

Conclusions

Adaptation processes are more than merely technical solutions to maintain or revert to the status quo. If adaptation processes can foster self-determined development and lead to transformation that ultimately allows marginalised groups to anticipate and prepare for change, adaptation to climate change for positive transformational change can be harnessed. Strengthening local institutions and development and engaging in monitoring can create learning platforms, access to information and opportunities for collective critical reflection, that enable local communities to better anticipate climate variability and thus prepare for possible future change.

In facilitating empowering participatory learning processes that enhance and ultimately harness self-confidence of women, we contribute towards a more just society. When designing interventions, it is important to consider that not all women are equally marginalised and that some women have more access to power and assets than others. Creating a safe learning space for all participants can be challenging, especially when communities are already stressed and conflicts are frequent.

Building individual adaptive capacity is a (potentially slow) process of personal development and will not be achieved in a project time frame of a few years. Commitment to these processes should be long term and from all parties: this is a slow learning journey towards transformative adaptation. It can be a personally rich and rewarding experience to participate in processes that expand networks and create innovative new pathways towards a more just future.

Table 16.2 Possible strategies for promoting the involvement of women in adaptation pathways for transformation

Challenge	Aim	Process/tools	Possible synergy effects
Women are often less confident than men & most vulnerable to stressors	**Increase self confidence of marginalised groups**	Ensure everyone is invited personally Spend time with people & listen PAR process inclusive & interactive Locally appropriate action	Increased networks Diverse income opportunities Men and women experiencing new ways of interacting with each other during workshops
Women are very busy and have many obligations	**Women have the opportunity to engage in adaptation discussions and PAR process**	Create a time and legitimate occasion for women to participate Arrange for adequate transport Parallel children's workshop Workshops are fun (and satisfy multiple needs)	Engaging children and youth Learn self-confidence from an early age Exploring gender dynamics, joint learning
Women often do not have wide networks (focus family networks)	**Expanding women's social and professional networks**	Engage in on and off farm research Create space for own ideas and experiments Support presenting experiences and findings at meetings and conferences	If carefully supported more self confidence Expanding networks could create new avenues of career and employment
Women are often shy to engage and share knowledge with a wider group	**Women freely engage in processes that interest them**	Create safe and appreciative spaces for sharing Work in smaller "groups of comfort" Ensure Intellectual property rights are protected	Increased self confidence Strengthening social cohesion in the community Calm spaces to engage with complexity
Women often are constrained in implementing their own ideas/experiments	**Women and men creatively explore and critically engage in experimentation and research**	Local PAR process Ensure the topics to explore are locally driven Explore joint research questions Create enabling frame-conditions for interested local researchers	One successful local woman scientist can be important role model Improved gender dynamics Expanded networks
Women are often hesitant to use technology	**Women use technology for exploring their own questions**	PGIS, PV, Water monitoring process, etc. Create safe spaces for exploring technology Work with interested women	Using technology has high status and is attractive Opening new career paths for young women

PAR processes for adaptation are delicate processes, and participation of men and women farmers and scientists in action research processes has been shown to create a deeper understanding of complex systems and an appreciation for all types of knowledge to address the challenges that humanity is currently facing.

References

Archer ERM (2006) Climate change in Namaqualand – Paper presented at the SKEP Conference. VanRhynsdorp

Archer ERM, Oettlé NM, Louw R, Tadross MA (2008) 'Farming on the edge' in arid western South Africa: climate change and agriculture in marginal environments. Geography 93(2):98–107

Babugura A (2010) Gender and climate change: South Africa case study. Heinrich Boell Foundation Southern Africa, Cape Town

Buenavista G, Flora CB (1994) Participatory methodologiesfor analyzing household activities, resources and benefits. In: Feldstein HS, Jiggins J (eds) Tools for the field – methodologies handbook for gender analysis in agriculture. Kumarian Press, Bloomfield

Chambers R (1983) Rural development – putting the last first. Longman Group, New York

Chambers R (2005) Participatory mapping and geographic information systems: whose map? Who is empowered and who disempowered? Who gains and who loses? Electron J Inf Syst Dev Countries 25(2):11

Engelbrecht F, Landman W (2010) Future climate: regional scenarios of future climate change over Southern Africa. In: South Africa. Dept. of Science and Technology (ed) South African risk and vulnerability atlas. Department of Science and Technology South Africa, Pretoria

Feldstein HS, Jiggins J (1994) Introduction. In: Feldstein HS, Jiggins J (eds) Tools for the field – methodologies handbook for gender analysis in agriculture. Kumarian Press, Bloomfield

IPCC (2012) A special report of working groups I and II of the intergovernmental panel on climate change – summary for policymakers. In: Field CB, Barros V, Stocker TF, Qin D, Dokken DJ, Ebi KL, Mastrandrea MD, Mach KJ, Plattner G-K, Allen SK, Tignor M, Midgley PM (eds) Managing the risks of extreme events and disasters to advance climate change adaptation. Cambridge University Press, Cambridge/New York, pp 1–19

Kemmis S, McTaggart R (2000) Participatory action research. In: Denzin D, Lincoln Y (eds) Handbook of qualitative research. Sage, London, pp 567–605

Koelle B, Oettlé NM (2009) Adapting with Enthusiasm: climate change adaptation in the context of participatory action research. Paper presented at the sustainable land use conference, Windhoek

Ledwith M, Springett J (2010) Participatory practice – community-based action for transformative change. The Policy Press, Portland

Lotter D (2010) Agriculture – an overview of agriculture in South Africa. In: South Africa. Dept. of Science and Technology (ed) South African risk and vulnerability atlas. Department of Science and Technology South Africa, Pretoria

Lunch C (2007) The most significant change: using participatory video for monitoring and evaluation. Participatory Learn Action 56:28–32

Oettlé NM (ed) (2005) Enhancing sustainable livelihoods in the Suid Bokkeveld. Both Ends, Amsterdam

Oettlé NM (ed) (2006) Knowledge and UNCCD: the community exchange and training programme. Ashgate, Burlington

Oettlé NM, Koelle B (2003) Capitalising on local knowledge: community knowledge exchange. The World Bank, Washington, DC

Oettlé NM, Arendse A, Koelle B, Van Der Poll A (2004) Community exchange and training in the Suid Bokkeveld: a UNCCD pilot project to enhance livelihoods and natural resource management. Environ Monit Assess 99(1–3):115–125

Oettlé NM, Goldberg K, Koelle B (2009) The Heiveld Co-operative – a vehicle for sustainable local development, Drynet case study. Drynet, Cape Town
Pelling M (2011) Adaptation to climate change – from resilience to transformation. Routledge, London/New York
Schipper L (2007) Climate change adaptation and development: exploring the linkages. Tyndall centre working paper 107, University of East Anglia
Schipper L, Burton I (eds) (2009) Understanding adaptation: origins, concepts, practice and policy. Earthscan, London
Senge P, Scharmer O (eds) (2001) Community action research: learning as a community of practitioners, consultants and researchers. Sage, London
Statistics South Africa (2003) Census 2001 in brief. Statistics South Africa, Pretoria
Sutherland A (1994) Managing bias – farmer selection. In: Feldstein HS, Jiggins J (eds) Tools for the field – methodologies handbook for gender analysis in agriculture. Kumarian Press, Bloomfield
Vogel C, Moserb SC, Kaspersonc RE, Dabelkod GD (2007) Linking vulnerability, adaptation, and resilience science to practice: pathways, players, and partnerships. Glob Environ Chang 17(3–4):349–364

Chapter 17
Gendered Adaptations to Climate Change: A Case Study from the Philippines

Gerlie T. Tatlonghari and Thelma R. Paris

Abstract Over recent decades, climate change has rapidly become a serious threat to human society and well being. One of the key identified effects of changing climate is the expected increase in flooding events. Flooding, brought about by heavy rainfall and frequent typhoons, is predicted to increase in a climate change prone country like the Philippines. As a country which often experiences climate-related disasters, men and women have developed adaptation strategies that make them resilient to extreme weather events. Using gender-sensitive qualitative methods, this study shows the gendered adaptation to flooding in rice farming communities in Nueva, Ecija Philippines. Based on the initial results of the study, men and women adapt to flooding according to their traditional roles. Moreover, due to changing weather patterns which greatly affect their livelihoods, men and women have assumed new roles. This alteration of gender roles has significant impacts on men and women farmers and on the welfare of their households. This study highlights the importance of understanding the gendered adaptation strategies and their consequences that can provide a basis for designing long term strategies for adaptations to climate variability.

Keywords Gender • Climate change • Adaptation • Philippines • Flooding

---

G.T. Tatlonghari, Ph.D. (✉)
Social Work Department, Gender and Social Sustainability Unit (GLASS),
Monash University, Melbourne, Australia
e-mail: gtat4@student.monash.edu

T.R. Paris, Ph.D.
Social Sciences Division, International Rice Research Institute (IRRI), Los Baños,
Laguna, Philippines
e-mail: t.paris@irri.org

M. Alston and K. Whittenbury (eds.), *Research, Action and Policy: Addressing the Gendered Impacts of Climate Change*, DOI 10.1007/978-94-007-5518-5_17,

Introduction

"Climate change is upon us" (Intergovernmental Panel on Climate Change 2007) and the effects of it are very threatening in a disaster-prone country like the Philippines. Recent frequent typhoons and unusual flooding events in the country are deemed to be a consequence of climate change (Allen 2006; Dioquino 2011). Thousands of people have been displaced, have lost their livelihoods or in some cases their lives. Climate change has grave consequences, a phenomenon described by the head of the Climate Change Commission of the Philippines "as a matter of life death for the Philippines" (GMA Network 2011). Disasters in the advent of climate change are likely to be more frequent unless people learn to adapt to its negative consequences.

Adaptation to climate change is about reducing vulnerability to observed or expected change in the climate and associated weather events (Adger et al. 2004, 2007; CARE 2010). However, the ability to cope and capacity to adapt to climate change varies across countries and social classes, where the poor and women will be hardest hit by climate change (IPCC 2007). Because of gendered norms, ascribed roles and inequalities in resources and power, women are more constrained compared with men to take action on climate change. Moreover, because of adherence to existing gender roles and relations, actions taken by women have increased their vulnerability during extreme climatic events. Several studies around the world have confirmed that climate change impacts are gendered and that the consequences of a changing climate are worse for women (Terry 2009; Demetriades and Esplen 2008; Alston 2007, 2010; Lambrou and Nelson 2010). Similar stories can be found in the Philippines as weather-related disasters such as flooding have increased women's vulnerability and domestic burdens as women are responsible for borrowing money, securing food in case of crop failure, caring for the sick, cleaning and maintaining their houses after flooding (Peralta 2009). Meanwhile, some studies also showed that because of some adaptation strategies undertaken by women during extreme weather events, there has been some alteration in the existing gender roles which have caused a shift in existing power relations within the household and community (Enarson 2000; Lambrou and Nelson 2010; Sultana 2010). These findings suggest that gender is a critical factor in exploring different adaptation strategies and understanding vulnerability to climate change.

To demonstrate how climate change could affect men and women differently, we present data from a case study we have conducted in rice farming communities in the Philippines that have been increasingly exposed to severe flooding. We argue that because of the gender roles and relations, men and women adapt differently and consequently face the impacts of flooding differently. We also highlight how men and women rose above the devastating situation which made them resilient to flooding. Finally, we stress the need to improve the adaptive capacity of men and women to reduce the negative impact of extreme weather events especially on women's lives.

Climate Change in the Philippines

The Philippines is ranked in the top 10 countries worldwide at risk for both climate change and disasters. According to a World Bank study (2010), during the last decade, the country experienced the highest recorded rainfall and the strongest typhoons. Weather-related disasters accounted for 98% of lives affected and 78% of lives lost between 2000 and 2008. For instance, the 2009 typhoons Ondoy and Pepeng alone had estimated damages and losses of US$4 billion, or almost 3% of gross national product. Although recent disasters cannot be totally attributed to climate change, recent information by the Philippine Climate Change Commission reported that a tropical country like the Philippines is expected to be increasingly exposed to sea level rise, extreme weather events and warmer days and nights (Dioquino 2011). Consequently, the Philippine government is now seriously acting on issues of climate change as "millions of Filipinos are already suffering, yet are only seeing initial climate change impacts" (GMA Network 2011).

Speculation about whether flooding is a consequence of climate change was recently discussed as more and more flooding occurs in the Philippines even in the areas where flooding is unknown. Though there are a lot of possible causes of flooding, climate change can cause flooding as a result of sea-level rise and an increase ocean and land temperatures. In the case of the Philippines for instance, these changes would mean more typhoons and variability of rainfall, therefore resulting in frequent drought and flooding. Studies conducted by Concepcion and colleagues (Concepcion et al. 2011) show that these phenomena are happening in different parts of the country, where farmers reported that they can no longer predict the shift of wet and dry seasons. They mentioned that there have been abrupt weather patterns, from very hot weather to short episodes of heavy rain showers all in 1 day. They also reported that from 2006 to 2009, typhoons arrived earlier and were of greater frequency and intensity.

Climate change is expected to warm the ocean and to enhance the strength, frequency and ranges of tropical cyclones. The enhanced potential of heavy rain combined with strong winds from more frequent typhoon crossings implies an increased risk of flooding. According to the Emergency Events Database (EM-DAT n.d.) produced by Centre for Research on the Epidemiology of Disasters (CRED), for a 100-year period (1910–2010), it was storms and floods that affected the greatest number of people in the Philippines. In terms of economic damage, flood caused the most economic damage for the last century, accounting for the loss of US$7.3 million in 1995, in one single flooding event.

The Study Area

Two rice farming communities in the town of San Antonio, Nueva Ecija were chosen as areas of the study. In order to understand how men and women deal with increasing flooding event, four focus group discussions (FGD), 13 key informant interviews and 21 in-depth interviews with three couples, 10 men and 8 women were conducted in 2011. Most of the farmers interviewed are above 40 years old and had been farming for at least 10 years in their communities. This enabled reflection on their experiences when faced with different kinds of flooding and climate conditions.

The province of Nueva Ecija is a landlocked province in the northern part of the Philippines. It is considered the main rice growing province and popularly known as the "Rice Granary of the Philippines". However, the province is ranked second in terms of areas susceptible to flooding in the country. To explore the extent of impacts of flooding in the lives of the men and women who rely heavily on rice farming, two villages in the town of San Antonio in Nueva Ecija were selected as the study sites. The town of San Antonio is considered the catch basin of the neighboring provinces such as Pangasinan, Tarlac and Pampanga where flood water traverses across before draining to Candaba Swamp in Pampanga. Of the 16 villages of San Antonio, 10 are considered low lying "barangays", or villages, and are prone to flash flooding and submergence. Barangay Papaya and Cama Juan are two of those areas susceptible to flooding. These villages are near or bounded by rivers or creeks thus, flooding is considered an ordinary event.

The village of Papaya has a total land area of 2,005.5 ha and has a total population of 6,766 with 1,572 households. Seventy nine percent (1,578 ha) of its total land area is devoted to rice farming of which more than 60% is vulnerable to flooding. Cama Juan, on the other hand, has only 510.66 ha of land and has a population of 2,258 and with 469 households. As Barangay Cama Juan is located in a more low-lying area, people living there consider flooding as more damaging to their livelihood and household welfare. Farmers describe flooding is part of their life. Flooding may start from the months of July to October which made it impossible for them to plant rice during the wet season. The dry season is their only time to plant and earn from rice farming and other farm related activities.

"Flooding is Becoming an Unpredictable Event"

Because of climate variability and changing rainfall patterns in recent years, flooding is now considered to be unpredictable. According to farmers, flooding depends on the number of typhoons and the rainfall intensity. They emphasized that there was an increasing number of floods in the last 10 years and flood occurrence varies each year. One of the farmers in focus group discussion described flooding as an unpredictable event.

> Flooding here is becoming an unpredictable event. It is like an oblivious old folk, it forgot when it should be coming (woman farmer 1 in FGD).

The average number of floods ranges from two to three during the wet season, usually in the months of July and September to November. With 2–3 days of heavy rains, fields may remain submerged for up to 14 days, while fields in very low lying areas are flooded for a month. Flooding not only devastates their fields but also their residential areas. If heavy rain occurs, people experience walking in chest-height flood waters particularly when large amounts of water are released from a dam. However, there was a recent flood in 2009 during the dry season, when farmers had just harvested their crops, which for them is a very unlikely event. Farmers recalled their experiences during that flood:

> We already threshed the paddy we harvested but it was washed away by floods – (woman farmer 1 in FGD)
>
> I cried when I saw our crop destroyed by the floods. I estimated that I should have harvested 250 cavans of palay [US$6,153 income equivalent] – (woman farmer, key informant).
>
> There was a time that we harvested our paddy from the floods. When you cut it from its stalks you need to embrace it and put it in the dry place. It's like you need to rescue somebody from dying. That was a tearful situation, it should have been our income already (male farmer 2).

Moreover, farmers mentioned that flooding is also caused by infrastructure problems. Participants in the study said that if there was a good dikes and floodway system, flooding could be minimized in their areas. Irrigation canals, particularly in Barangay Papaya, are not maintained properly which causes flooding in residential areas.

The Gendered Consequence of Floods

Men and women are both affected by floods in different ways because they have different roles and resources, including control and power, which are shaped by cultural rules and norms (CARE 2010). Therefore their experiences are varied, which should be given due attention when investigating the social impacts of extreme weather events. Men are expected to increase their workload, before, during and after flooding as they need to save their crops, and adjust their farming activities, if needed, to avoid the coming of the floods. Women, on the other hand, increase their time in preparing food and securing their house and livestock when typhoons are about to come at the same time securing money for the needs of the farm and the household. Crop loss, increased debts, hunger, isolation and weakening of health were identified by both men and women as the major negative effects of flooding.

The loss of livelihood due to flooding means people need to secure cash income from other sources aside from rice farming. On the other hand, many households continue to plant rice and take the risks of losing their crop due to anticipated floods. Consequently, those households without enough capital outlay are trapped in a cycle of debt. They borrow money during the planting season, which they repay during the havesting season and then acquire another loan for the next planting season. Falling into the debt trap is a direct result of frequent flooding. However, more women expressed this as a burden as it is women who manage the finances of the farm and budget household expenses. One farmer, a widower, described his loan as "chains" that do not end.

A woman farm manager related her experience in borrowing money.

> My worst experience was when the money lender took away all the paddy which I harvested because I couldn't pay my loan. I cried and cried due to lost income. In this village the interest rate is high. If you borrow 8 cavans [456 kilos equivalent to US$197], you need to pay the money lender 3 times (the borrowed amount). Thus, I also have to bear the responsibility of repaying my loan and to find other sources of income, aside from farm income. So I told myself, I need to have an additional income just to pass through this (key informant and a woman farm manager).

However, there are situations when husbands do not include their wives in making decisions related to farming and investments. These husbands shoulder all the burdens as expressed by a sole male provider of the family.

> My wife doesn't work in our farm. I don't like our children to see my wife doing hard labour in the farm. I let her do all the household chores and child care responsibilities. So I do all the work in the farm and take full responsibility for looking for money to be used in farming. I think I am a survivor (...). I don't have relatives and I have no one to go to for help (male farmer 4).

Women's role in securing loans from private money lenders does not mean that they are empowered. In fact, they bear the stress or pressure to make ends meet while, at the same time endeavouring to secure the health and nutrition of their children. One woman farmer lamented:

> Due to floods, we did not harvest anything. Seeing our crops ready to be harvested being swept by floods and disappearing was too painful for me (woman farmer 1 in FGD).

Another woman in the FGD shared her experience of food is not being available during floods:

> All the families are very sad because we don't have anything to eat. You are fortunate if you have stock food in your house but if you don't have, you are pitiful. You just have to survive by drinking warm water (woman farmer 2 in FGD).

Moreover, flooded roads also restrict women's movement such as buying or looking for food to eat. Although floods lead to crop loss and less access to food, women rely heavily on their social networks. They said that they can easily ask their neighbours to lend them some rice and most of their neighbours share food e.g. fish and vegetables especially during tough times. Men, on the other hand, suffered from physical isolation and worried more about crop loss due to floods. They also felt useless as they could not do anything while waiting for the floods to subside. Some of them dared to walk in the floods which led to health problems for example athlete's foot and rheumatism in men. Flooding also brought more emotional stress on men than on women.

In one interview with a husband and wife, the woman expressed her concern about her husband's health and emotional condition:

> My husband just had a mild stroke this year. When we lost our crops due to flooding, my husband used to go out drinking to forget our loss and unpaid debts (wife 1).

The husband on the other hand, expressed how his wife endured during crop loss.

> My wife doesn't seem to have a problem whenever we lose our crops. She seems happy every day (husband 1).

However, the woman disagreed with the husband's opinion, expressing that she also feels stressed about their situation, but she just stood strong to handle their loss. This suggests that more often, men are more emotionally affected by the stress due to crop loss while women are more resilient during tough times. Women are more receptive to take on behavioural changes than men, as more women believe that environmental problems cannot be solved without changing lifestyle (Brody et al. 2008: 15).

Gendered Adaptation to Flooding

It is clear from the previous discussion that impacts of flooding are gendered, as men perceived and responded quite differently to women. It is observed that men's and women's varying responses are often dictated by their ascribed roles, gendered norms and relations. Moreover, there are is evidence to show that alteration of gender roles and relations is another way of adapting to the effects of frequent flooding.

Men and Women Respond Based on Their Traditional Roles in the Household and Farming

In Filipino society, men are culturally perceived as major income earners and "farmers" in the rural areas. On the other, women assume the roles of mothers and custodians of household finances. Although women work as unpaid family labour in transplanting, weeding, harvesting and threshing, their roles as farmers are unrecognized and underestimated. Thus they are often referred to as "housewives" and excluded in agricultural extension programs. However, through time, farmers learned to adapt and adjust their farming system, especially when they expect typhoons/floods to occur two or three times in a year. Based on the focus group discussions and in-depth interviews with male farmers, they mentioned the following adaptation strategies:

- adjust rice planting: ahead or late based on the occurrence of flooding;
- resow when seedlings are destroyed;
- replant when there are available seedlings for transplanting;
- buy seedlings from other farmers;
- choose varieties with high yields and tolerance to flooding;

- build higher dikes around the plot;
- secure crop insurance;
- store all farm machinery in secure places;
- delay harvesting and
- harvest whatever could be saved after flooding.

On the other hand, women tend to be more creative and more resourceful in using their social network for borrowing money, food and also other support services. Some of the strategies which women do during flooding are:

- stock food before and during flooding season;
- spend less on unimportant commodities;
- secure important assets and livestock by protecting them from floods;
- take care of the nutrition and health concerns of the family members;
- seek assistance from the government (relief goods) and
- secure loans for household needs and additional farming expenses from relatives and friends.

Women Learning to Adjust Their Roles in Farming and Households

One of the major effects of seasonal flooding is the loss of livelihood opportunities in rice farming. Since men and women know that flooding usually occurs during the wet season, they maximize their opportunity to earn during the dry season and save their earnings for the rainy days. Men also recognise the need to involve women in on-farm and non-farm activities, which would allow them to contribute to household income. Consequently, women are increasingly engaging in different productive activities especially in rice farming. Their involvement in productive activities not only opens new roles and opportunities for women but becomes one of the major adaptation strategies of the farming household during flooding. Women take on new roles as they engage into productive activities.

From Farmer's Wife to Farm Manager

Due to threat of crop loss due to floods, many women began to assume managerial responsibilities. Some of the women interviewed said that they convinced their husbands not to sell or mortgage their lands. Land is one of the major household assets which farmers are not willing to forego, even in tough times. Women farmers pointed out that there are several factors that paved their way to become farm managers.

Mechanization

Traditionally, the principal male ploughs the land with the use of water buffalo (carabao) while the principal female helps clean the dikes. However, with mechanization (use of tractor and hand power tiller), the principal female no longer deals with arduous and dirty jobs but instead, supervises hired workers in land preparation.

Membership in Farmers' Association

Traditionally membership in farmers' associations is exclusive to men. However, due to the strong advocacy for empowering women and reducing gender inequalities in access to resources by local governments and non-government organizations, more women farmers are now officially accepted as members in farmer organizations. Female memberships in farmers' organization automatically give them access to training, credit and farm inputs. Moreover, principal women are now allowed to substitute or represent the principal men if they cannot attend meetings due to their busy schedule in the farm. Women who have better access to training and credit through membership in farmer associations are more empowered to manage farms.

Long Term Experience in Farming

Most principal women were engaged in rice farming when they were young and unmarried. Thus they are repositories of indigenous knowledge and skills which are required in managing farms efficiently. As an active member of the household, community leader or member, steward of natural resources, women can offer different resources, indigenous knowledge and understanding of the causes and consequences that could significantly contribute in responding to extreme climatic events (Peralta 2009:18).

Education

Research shows that in the Philippines, land is preferentially given to sons, while schooling is given to daughters (Estudillo et al. 2001). Parents invest more in daughters' schooling because returns to female schooling have risen in the non-farm sector while land is given to land sons because rice farming is perceived as involving male labour. On average, women farmers who participated in our study have completed 6 years in elementary school or dropped out of high school. The better educated women farmers are more capable of managing their farms, despite their withdrawal from field work. Furthermore, educated women who took over managing the farm are more willing to take risks and to experiment and participate in adaptive research activities because they can easily understand the concepts behind complex technological innovations.

From Budget Keepers to Entrepreneurs

Due to climate change, women bear the burden of allocating or stretching the limited budget for different expenditure and saving in anticipation for emergencies. Thus, they seek different sources of cash income such as fattening swine, raising poultry, growing vegetables for consumption and sale in the village, engaging in small trade e.g. buying and selling rice, or acquiring land through mortgaging ("sanglaan") and other expenditure savings. They are increasingly becoming entrepreneurs using agricultural products. However, this is not possible without the approval of their husbands. Female heads of households (widows and separated women) tend to be more entrepreneurial, or under more pressure to be creative in finding alternative sources of cash.

With new roles these women have taken, they gain self-worth and self confidence. A woman farm manager shared her experience with pride.

> I applied what I have learned in farmers' meetings and training activities in managing our 11 hectare farm land, which we mortgaged. However, in just one harvest, I was able to recover our fields we were forced to mortgage. As a consequence my husband has learned to recognize my skills. He told me 'Abay [referred to as partner] I entrust to you our farm land' and I was able to prove to them [husband's relatives] that I am a farmer (key informant and a women farm manager).

Flooding brought about changes in roles of men and women in many rice farming households. These changes influenced decision-making and consequently livelihood security. As women are getting more involved in farm management most husbands and wives agree that they should participate jointly in making decisions related to farming and investments. Some of these major decisions were:

- to plant or not plant during wet season;
- where and how much to loan as capital for rice farming;
- what other non-farm income sources they should engage in and
- how much rice to store for food consumption for the whole year.

> We both make decisions on matters related to farming and our business. We do not let one dominate or not be informed about the decision made (interview with husband and wife).

Involving women farmers in decision-making allowed their voices to be heard. Moreover, they gained more access to and some control over resources that were not available earlier. Through their participation in the agricultural training programs, they had more access to seeds, fertilizer and technical knowledge. In fact, women's influence now extends beyond household decisions to decisions related to budget allocation for farm input expenses and productive activities (Bernabe and Cantos 2007). If women are given equal opportunities with men, they can help in better farm management, in particular, reducing farm expenses. As shared by a wife who has been actively managing the farm with her husband:

> In our household, we have specific roles and division of labour. I monitor our farming expenses, buy the fertilizer and choose what varieties to grow. My husband does all the field work. I share my knowledge and skills I gained from the training with him. One of the benefits of both husband and wife working together is being able to save money by not hiring labourers (wife 2).

Positive Outlook About Rice Farming

Although flooding has an adverse effect on the household welfare, particularly on rice farming, positive attitudes to climate variability and rice farming influence the way farmers adapt to flooding. Flooding, for most farmers, is a natural event that cannot be stopped. Their long experience of flooding made farmers accept that it is already part of their lives. Moreover, Filipinos are very religious people. They cling to their faith and attribute God as the controller and provider of everything. Thus, they are more accepting of their loss especially in rice farming.

> I willingly accept it (loss from flooding) though it is painful; I cannot do anything about it (male farmer 10)
>
> We have so many prayers. If we could only block the floods, we will do that, but it's impossible to block it. We raise all our problems to God (woman farmer 1 in FGD).

It is surprising to know that despite the frequent floods, there are farmers who have a positive outlook in farming as the main source of their livelihoods. They attribute their positive outlook to technologies such as mechanization and high yielding varieties and the increased participation of women in rice farming. Most farmers are proud of their occupation despite its low status in Filipino society. They said that they will never give up farming and their farm lands. They are optimistic that they can improve their lives through rice farming and aspire to bequeath their farming experience and their lands to their children.

Strategies for Better Adaptation

Men and women are aware of the risk and uncertainty of the climate conditions and the future of their farming activities and thus expressed their need for increasing their capacity to secure and sustain their livelihoods. In order to effectively respond to flooding, men and women suggested the following strategies:

(a) Provide more accurate weather forecast to rural areas. It is important for farmers to have accurate weather forecasts as rice farming is fully dependent on climate. With accurate weather forecasts, they can reduce the anticipated negative effects of floods by adjusting their cropping calendar or growing short duration varieties that can enable them to escape or avoid possible typhoons and flooding.

(b) Introduce rice varieties for varied types of floods e.g. deep and stagnant flooding. For example, while the International Rice Research Institute (IRRI) and Philippine Rice Research Institute (PhilRice) have developed and introduced rice varieties which are tolerant to submergence (14 days), farmers still look for varieties which are not only tolerant to stagnant and deep flooding but which also have higher yields than the varieties they use, command good price and have good eating qualities. IRRI has introduced stress tolerant varieties (e.g. IR64 Sub1) in submergence prone areas in Central Luzon, Philippines.
(c) Provide women access to credit and training. More women expressed their need to have access to formal credit at lower interest rates. They also want to be trained on all aspects of rice production until post harvest to improve their yields.
(d) Provide better irrigation canals and floodway system that would regulate the flow of water to their rice fields.
(e) For the government to regulate the prices of farm inputs and fuel. This will enable farmers to increase their profits and encourage more farmers to continuously cultivate their lands.
(f) Provide alternative income generating activities. To supplement their rice farming expenses, farmers require other livelihood activities that are suitable to their environment. Women in particular would like to have other livelihood activities during the wet season when rice farming is not possible. They want income generating activities that can be undertaken in their homes or near their homes. Proximity is still a major concern to women as they still have the responsibilities of taking care of the household chores and child care. Men also like to engage in other livelihood activities but they are constrained by lack of capital and lack of required skills. Vegetable farming and livestock raising (e.g. duck raising) are potential source of income.
(g) Key agricultural agencies need to coordinate their activities to address the impacts of climate change. This is to avoid duplication in activities and to accelerate the dissemination of technologies and information to both men and women farmers.
(h) Commitment from stakeholders to carry out and sustain good projects and new technologies that farmers can use as part of their adaptation strategies.

Conclusion

The rice farmers interviewed strongly claim that climate change is now a reality on the ground. Increasing flooding events in the rice farming communities studied brought about by increasing typhoons and heavy rainfall have made these communities more exposed to climate hazards. These kinds of extreme weather events have resulted in loss of livelihoods and have escalated debt burdens that have long term welfare consequences among farming households. It is likely that same rice farmers will be trapped in this kind of this situation in the advent of climate change. However, this study also illustrates how men and women can be resilient in risky environments. They tend to continue in their gender roles, and to adhere to gender norms and relations as they adapt to flooding. However, prolonged exposures to flooding and continuous loss of livelihood have made some men and women realise the need to alter their gender roles and relations. These gender role changes have resulted in increased access to resources and empowerment among women who take on additional responsibilities and change their roles in rice farming and in the household. However, this has also brought additional burdens for women as they become more concerned in managing their farms and at the same time taking care of household welfare. Women's potential talents as key agents of change and their resourcefulness in adapting to climate change should be enhanced.

Women and men in the rice farming communities of San Antonio, Nueva Ecija, Philippines are good example of people who rose above an extreme event. Their different responses to flooding events prove their resilience to extreme weather events. However, their resilience can be diminished or lost if their resources continue to deteriorate and if there is an increasing uncertainty about climate conditions. Climate change will definitely challenge the existing farming conditions and the resilience of the people who are continuously exposed.

Although men and women can do much in order to reduce their vulnerability to climate change, there is a great need to have gender-sensitive policies and programs that can actually address the predicted impacts of climate change. Social policy must be given enough attention and careful planning to increase men's and women's resilience to climate change. Moreover, there is a need for the government, development organisations, research institutions, civil societies and communities to act collaboratively in order to maximise recent information about the changing climate and to efficiently utilise their existing resources to combat the negative impacts of climate change. Coordination among stakeholders is needed for a faster and more sustainable adaptation among individuals and communities.

References

Adger WN, Brooks N, Bentham G, Agnew M, Ericksen S (2004) New indicators of vulnerability and adaptive capacity. Tyndall Centre for Climate Change Research technical report No. 7. www.adaptationlearning.net/system/files/Tyndall.pdf. Retrieved 15 Dec 2011

Adger WN, Agrawala S, Mirza MMQ, Conden C, O'Brien K, Pulhin J, Pulwarty R, Smit B, Takahashi K (2007) Assessment of adaptation practices, options, constraints and capacity. In: Parry ML, Canziani OF, Palutikof JP, van der Linden PJ, Hanson CE (eds) Climate change 2007: impacts, adaptation and vulnerability contribution of working group II to the fourth assessment report of the intergovernmental panel on climate change. Cambridge University Press, Cambridge, pp 717–743

Allen KM (2006) Community-based disaster preparedness and climate adaptation: local capacity-building in the Philippines. Disasters 30(1):81–101

Alston M (2007) Gender and climate change: variable adaptations of women and men. Just Policy 46:29–35

Alston M (2010) Gender and climate change in Australia. J Sociol 47(1):53–70. http://jos.sagepub.com/content/early/2010/10/09/1440783310376848. Retrieved 16 Dec 2010

Bernabe R, Cantos J (2007) Gender and rice: the case of the Philippines for IGTN. In: Speldoc A (ed) The gender impacts of trade liberalization on our food system, agricultural markets and women's human rights. Friedrich-Ebert-Stiftun, Switzerland, pp 19–20

Brody A, Demetriades J, Esplen E (2008) Gender and climate change: mapping the linkages. A scoping study on knowledge and gaps. www.bridge.ids.ac.uk/reports/Climate_Change_DFID.pdf. Retrieved 4 Jun 2010

CARE (2010) Adaptation, gender and women's empowerment. http://www.careclimatechange.org/files/adaptation/CARE_Gender_Brief_Oct2010.pdf. Retrieved 10 Oct 2011

Concepcion RN, Espaldon MVO, Lewicki-Dhainaut SD, Dedicatoria RMM, Abucay ER (2011) Coping with extreme climatic events: stories of resiliency in the Philippines. Agriculture and development notes on climate change adaptation, vol 1, no 4. http://beta.searca.org/kc3/index.php/k-showcases/800-coping-with-extreme-climatic-events-stories-of-resiliency-in-the-philippines. Retrieved 10 Oct 2011

Demetriades J, Esplen E (2008) The gender dimensions of poverty and climate change adaptation. IDS Bull 39(4):24–31

Dioquino R-AJ (2011) Gov't official: climate change partly to blame for floods. GMA News. http://www.gmanews.tv/story/234397/nation/govt-official-climate-change-partly-to-blame-for-floods. Retrieved 7 Oct 2011

EM-DAT International Disaster Database (n.d.) http://www.emdat.be/. Retrieved 30 Aug 2011

Enarson E (2000) Gender and natural disasters. In: Focus programme on crisis response and reconstruction working paper 1. http://www.ilo.int/wcmsp5/groups/public/---ed_emp/---emp_ent/---ifp_crisis/documents/publication/wcms_116391.pdf. Retrieved 20 May 2010

Estudillo J, Quisumbing A, Otsuka K (2001) Gender differences in land inheritance, schooling and life time income; evidence from the rural Philippines. J Dev Stud 37(4):23–48

GMA Network (2011) Climate change a matter of life and death – PHL Commissioner. http://www.gmanews.tv/story/234435/nation/climate-change-a-matter-of-life-and-death-phl-commissioner. Retrieved 7 Oct 2011

IPCC (2007) Fourth assessment report (synthesis report). Inter-governmental panel on climate change

Lambrou Y, Nelson S (2010) Farmers in a changing climate does gender matter? http://www.fao.org/docrep/013/i1721e/i1721e00.htm. Retrieved 5 Nov 2010

Peralta A (2009) Financing for climate change mitigation and adaptation in the Philippines: a pro-poor and gender sensitive perspective. Reality check. http://www.realityofaid.org/userfiles/publications/roareport_6badc3e72d.pdf. Retrieved 10 Jan 2012

Sultana F (2010) Living in hazardous waterscapes: gendered vulnerabilities and experiences of floods and disasters. Environ Hazards 9(1):43–53

Terry G (2009) No climate justice without gender justice: an overview of the issues. Gend Dev 17(1):5–18

Worldbank (2010) A strategic approach to climate change in the Philippines. http://siteresources.worldbank.org/INTEAPREGTOPENVIRONMENT/Resources/PHCCSNJan27final.pdf. Retrieved 11 May 2010

Chapter 18
Gender and Declining Fisheries in Lobitos, Perú: Beyond *Pescador* and *Ama De Casa*

Naomi Joy Godden

Abstract Climate change affects the capacity of artisanal fisherfolk in Perú to generate a reliable income. Men's difficulty in fulfilling their traditional gender role as the primary income earner threatens social constructions of masculinity. International research explores the growing phenomenon of women engaging in non-traditional livelihoods for financial survival due to climate change. The case study of Lobitos, an isolated fishing village in northern Perú, shows that women experience emancipation and oppression with declining fisheries and livelihood adaptation. Economic and environmental uncertainty is connected to a departure from the gendered work roles of pescador [fisherman] and ama de casa [female household manager], vulnerability and safety concerns, enhanced focus on education, and increased female employment. This case study demonstrates that the machismo/marianismo paradigm, a prevailing dichotomous gender framework in Latin American literature, fails to recognise the multiplicity of gender identities and the processes of gender renegotiation in a community affected by climate change.

Keywords Gender • Climate change • Peru • Declining fisheries • Gender roles

Naomi Joy Godden Naomi is a Social Worker from Margaret River, Western Australia. She has over 10 years of experience as a professional and volunteer in rural community development, rural social justice advocacy, and social research. She has worked in small communities around Australia and internationally and has worked at Oxfam Australia. She is currently studying her Ph.D. in rural community work at the Gender, Leadership and Social Sustainability Research Unit at Monash University.

N.J. Godden, BSW (Hons), BA (Australian Indigenous Studies) (✉)
Gender, Leadership and Social Sustainability Research Unit, Department of Social Work, Monash University, Caulfield, VIC, Australia
e-mail: naomi.godden@monash.edu

M. Alston and K. Whittenbury (eds.), *Research, Action and Policy: Addressing the Gendered Impacts of Climate Change*, DOI 10.1007/978-94-007-5518-5_18,

Introduction

Adaptation is the "adjustment of a system to moderate the impacts of climate change, to take advantages of new opportunities or to cope with the consequences" (Adger et al. 2003, p. 192). Livelihood adaptation, particularly in natural resource-dependent communities, is an important factor in climate change adaptation (Lambrou and Grazia 2006; Thomas and Twyman 2005). Emerging international literature shows that gender(ed) roles in rural livelihoods are changing due to climate change (see Brandth and Haugen 2010; Alston and Witney-Soanes 2008; Baas and Ramasamy 2008; Mitchell et al. 2007). In coastal Perú, the lived reality of decreased and insecure fish stocks and growth of alternative livelihoods threatens the traditional coastal gender dichotomy of *ama de casa* [female household manager] working (unwaged) in the home and *pescador* [fisherman] working as the sole income generator. This has implications for gender identities and experiences, and the vulnerability, safety, education, and livelihoods of women.

This chapter explores women's experiences of livelihood adaptation due to declining fisheries in Lobitos, an isolated coastal village located 963 km north of Lima, Perú. I discuss the theoretical context of gender and climate change in Perú, and then analyse data from a 16-month research study in Lobitos, embedded with external research and women's (de-identified) stories. I argue that: (1) a dichotomous construction of gender is incompatible with the localised reality of declining fisheries; and (2) a renegotiation of gender roles and livelihood in Lobitos can both emancipate and oppress women.

Gender and Climate Change in Perú

Gender in Perú: "Machismo" and "Marianismo"

Socially constructed gender roles are evident throughout Latin America (Olivas-Lujan et al. 2008; Chant and Craske 2003). Despite improvements regarding the status of women, Latino societies remain highly patriarchal, reinforced by institutions such as the State, capitalism, and the Catholic Church (Peña 1995; Safa 1990). A prevailing theoretical framework suggests that gender in Latin America is dichotomised through *machismo* and *marianismo*. Stevens (1973b) defines *machismo* as the stereotyped "cult of virility" of the male: "exaggerated aggressiveness and intransigence in male-to-male interpersonal relationships and arrogance and sexual aggression in male-to-female relationships" (p. 315). *Machismo* encompasses sexist attitudes and behaviours through macho identity, displayed with self-confidence, sexual prowess, zest for action, bravery, and domination of women (Basham 1976). Through this lens, "the home" is considered the woman's natural place as a familial binding force as mother and wife. The terminology and meanings of *machismo* persist; for example, women in the popular sector of

Lima recognised *machismo* as the cause of many social wrongs, including domestic violence, men's resistance to family planning, opposition to women's involvement in other activities, possessiveness of women, and discouragement of female self-realization (Padilla 2004).

To complement *machismo,* Stevens proposes *marianismo* as the cult of feminine spiritual superiority, where women are "semi divine, morally superior and spiritually stronger than men" (1973a, pp. 90–91). *Marianismo* conceptualises Latina women as submissive, dependent and sexually faithful to their husbands, dedicating themselves to the household needs and their husbands and children (Flake 2005). Chant and Craske (2003) explain that the idealised femininity of women with spiritual and moral superiority to men 'legitimates' their subordinate domestic and societal role.

The stereotypical dichotomy of *machismo* and *marianismo* typifies Latino men as strong, brave, and confident, and Latina women as spiritual, caring, and compassionate. This binary structure places men working (waged) in public as the sole breadwinner, and women working (unwaged) in the private home. The *machismo/marianismo* dichotomy has been extensively critiqued as being exaggerated, simplistic, stereotypical, as victimising women, and negatively portraying men (see, for example, Beattie 2002; Fuller 2001; Mirande 1997; Hays-Mitchell 1995; Ehlers 1991). Alcalde (2010) argues that the hegemonic construction of masculinity emphasises dominance over women. Based on male primacy and subsequent right to control women, *machismo* has been construed to legitimise poor treatment of women (see Chant and Craske 2003). It also connotes exaggerated masculinity, male chauvinism and an "extreme male supremacist ideology" (Mirande 1997, p. 149).

Indeed, gender identities are not static and cannot be generalised. They are constituted within a multiplicity of differences and contexts, including age, class, ethnicity, race, regional, institutional and other categories, with numerous masculinities and femininities (Fuller 2001; Mirande 1997). In her study with Perúvian middle-class, urban men, Fuller (2001) identified three categories of masculinity: the natural (sexuality and physical strength); the domestic (virility and reproductivity through family, marriage, and fatherhood); and the 'outside'/public ('manliness' through work and politics). Padilla (2004) also challenges the construction of women's femininity in domestic and reproductive domains, highlighting women's desire for egalitarianism and participation in social action for change.

Despite scholarship challenging the *machismo* and *marianismo* dichotomy, it is indisputable that the socio-economic position of women in Perú is low. In 2010, the United Nations Development Program (UNDP) ranked Perú at 63 of 169 countries on the Human Development Index, and 74 of 138 countries on the Gender Inequality Index. Perúvian girls and women experience serious concerns in areas such as maternal morality, education, employment, and domestic violence (UNDP 2010; Instituto Nacional de Estadistica e Informatica (INEI) 2010; Garcia-Moreno et al. 2005).

Furthermore, distinct livelihood roles of Perúvian women and men exist. Women generally engage in household work while men are involved in income generation (women also work one-fifth more than men) (Ilahi 2001b). The private home is the woman's domain while the public sphere "sustains and legitimizes masculine predominance" (Fuller 2001, p. 319; see also Alcalde 2010). Women are expected to maintain the household, including supporting her husband's physical, emotional and sexual needs (Alcalde 2010), and engaging in motherhood, a concept widely venerated in Latin America (Chant and Craske 2003). For men, work is a "prerequisite for establishing a family and the principal source of social recognition" (Fuller 2001, p. 319), and guarantees their dominance as responsible providers. However, Latin America, including Perú, is undertaking radical social change through women's participation in the paid workforce (Olivas-Lujan et al. 2008). As discussed later, climate change contributes to this phenomenon.

Climate Change in Perú: the Future of Fishing

The Intergovernmental Panel on Climate Change (IPCC) (2007) presents numerous predictions for Latin America and Perú regarding the coastal impacts of climate change, including: sea-level rise which will impact fisheries, coastal inundation and erosion; coastal storm regime modification; altered coastal morphology; disrupted access to fishing grounds; negative impacts on biodiversity; pollution of marine environments; intensified El Niño-Southern Oscillation (ENSO) events; and increased wind stress, hypoxia and deepening of thermocline which will impact the marine ecosystem and fisheries (e.g. reduced spawning areas). The predicted socio-economic and health-related impacts in coastal areas include: population displacement; reduced access to drinking water; negative impacts on coastal tourism; water-borne epidemics due to flooding; reduced fisheries; significant health concerns, exacerbated by migration; food insecurity and others. Previously, the north-west region of Perú (where Lobitos is located) has experienced serious floods due to ENSO variability, a natural 3–8 year cycle with significant socio-economic impacts. ENSO events are difficult to predict and highly varied (Young and Leon 2010), and although they are expected to intensify due climate change, the actual relationship between climate change and El Niño is still unclear (Sperling et al. 2008).

Previous research shows that fisheries are highly volatile in Peru (see Pauly et al. 1998), due to varying fish stocks, overharvesting of fish (Ibarra et al. 2000), market demand, and climate change (IPCC 2007). Fisheries contribute between 0.8 and 1.4% of the Peruvian Gross Domestic Product, and artisanal (small-scale) fishing provides employment, poverty mitigation, and food for nearly 38,000 artisanal fishers and their families on the Perúvian coast (Food and Agriculture Organisation (FAO) 2010). The Perú-Humboldt Current System on the western coast of South America provides the world's largest pelagic fisheries, but is considered a volatile eco-system due to climate risks (Bertrand et al. 2004). These risks include changes to the marine environments, fish stocks, fish movement, ecosystems and coastal communities, ocean acidification due to increased carbon dioxide, and extreme weather events (Daw et al. 2009; IPCC 2007). The documented decline of Peruvian fisheries (Pauly et al. 1998) reinforces the need for fisheries adaptation to reduce climate change vulnerability and associated fisherfolk poverty (Badjeck et al. 2010).

Gender and Declining Fisheries in Lobitos

This section considers the convergence of gender and declining fisheries in the community of Lobitos, Perú.

Community Context: Lobitos

Lobitos is a coastal community of 1,506 people (920 males and 586 females (INEI 2007)), located in the Piura region, northern Perú. In the 2007 Peruvian census, 419 men and 114 women above the age of 6 years were considered employed or employable, of whom 94 men (22%) and 30 women (26%) had difficulties generating income (INEI 2007). The main waged activities for men included: artisanal fishing (148 men; 37%), real estate and renting (67 men; 16.5%); public administration and defence (54 men; 13.3%); and construction (32 men; 7.9%). The main waged activities for women included: commercial businesses (48 women; 42.8%); hotels and restaurants (15 women; 13.4%); real estate and renting (14 women; 12.5 %); public administration and defence (11 women; 9.8%); other work such as social, community and personal services (11 women; 9.8%); and domestic workers in private homes (10 women; 8.9%). The rate of female employment in Lobitos (22%) was significantly lower than the regional average; in 2007, 66.2% of Piuran women had worked in the previous 2 months (INEI 2010). The census information does not adequately reflect the temporality of most work in Lobitos, including the short-term casual roles in construction and local government, nor the estimated 40 people (generally fishermen) who provided casual transport services with their *moto-taxi*, or the 30–50 homes with a small shop or business for additional income. Furthermore, these statistics do not show whether workers were vulnerably employed, as 39.6% of Perúvians are vulnerably employed, with a ratio of 1.41 females to every male (UNDP 2010).

No formal research has been conducted regarding climate change in Lobitos; however, local fisherfolk narratives report several decades of a continued, significant decline in fish resources, and serious climate events. Lobitos has an arid climate with very little annual rainfall, but has high risk of El Niño flooding, which has previously caused significant damage to homes and infrastructure, food insecurity, and income generation difficulties (see Reyes 2002). The 1983 El Niño flood pushed a significant amount of sand into the local beach, forming a surfing wave. The 1998 El Niño flood enhanced the sandbanks and wave quality, triggering surf tourism. Lobitos is considered a world-class surfing destination and has hosted several international surfing competitions.

According to local narratives, between 2000 and 2010, surf tourism in Lobitos grew from a single hostel to more than 20 accommodation providers and 10 restaurants, with 50–100 Perúvian and foreign tourists visiting at any given time. Nearly all tourism services were owned and managed by approximately 50 *turist locales* [tourist locals] – middle-class people who moved from Perúvian cities or overseas. Surf tourists spent up to $50 US per day for food and accommodation, 3–20 times the daily fishing wage. However, very few local people were employed or financially benefitted from tourism. Other localised industries include a declining military base of 150 soldiers, and several international petroleum companies with onshore and offshore drilling; employing few, if any, local people.

Between July 2009 and October 2010, I was the Program Manager of the non-government organisation WAVES for Development (WAVES) in Lobitos. WAVES "creates life-enriching experiences in coastal communities through Education programs that develop youth into healthy & empowered adults and Surf Voluntourism programs that engage travelers & transform their views of the world and themselves" (WAVES 2012). The acronym WAVES stands for (summarised from WAVES 2012):

Water: Celebrating the relationship between people and water;
Adventure: Building confidence, staying healthy and fostering meaningful relationships through adventure;
Voluntourism: Combining volunteering and tourism;
Education: Creating education opportunities for young people;
Sustainability: Enhancing ecological, social, political and economic sustainability.

In Lobitos, WAVES employs ten local and international staff members who manage education and volunteer programs to engage local young people and international travellers.

Research Methodology

During 16-months in Lobitos, local workers, community members, and I conducted research to understand local socio-economic relations and to inform the WAVES program. The descriptive research considered the actions, events, beliefs, attitudes, and social structures and processes occurring in the community (Marshall and Rossman 2011). Nestled within feminist and cultural studies theoretical paradigms, the research design had a grounded theory strategy (Corbin and Strauss 2008), utilising mixed methodologies of data collection. Methods included a household survey with quantitative demographic questions and qualitative semi-structured interviews, focus groups, and participant observation. In accordance with participant observation principles, my direct community involvement enabled me to observe and experience interactions as an insider (Marshall and Rossman 2011; Angrosino 2008; Jorgensen 1989). Jorgensen argues that the participant observer should play multiple roles during the project, gaining rapport and even intimacy with the people, situations, and settings of the research.

The household survey was the key data source, and included semi-structured interviews with adults in 50 homes (57 females and 31 men) in the main *barrio* [neighbourhood]. Discussion themes included: local issues and strengths; perceptions of WAVES; ideas for WAVES educational surf program and community projects; and tourism. Quantitative inquiry included household member characteristics of age, sex, occupation, and literacy level. Two focus groups included parents and WAVES students. Importantly, the household survey and focus groups had an emancipatory focus, engaging and training local people in data collection, analysis, and report writing. Interviews were always conducted by a local woman and me in partnership, for cultural sensitivity and trust-building. Additionally, field notes from casual conversations, observations of local rituals, practices, tourism, and WAVES programs documented the socio-economic and gender context.

Livelihood Adaptation: Beyond "Pescador" and "Ama de Casa"

Local employment statistics and initial community observations suggest that gender livelihood roles in Lobitos reflected the *machismo/marianismo* scenario. Certainly, the common primary professions amongst the working-aged household survey respondents were *pescador* [fisherman] (42 men) and *ama de casa* [female household manager] (48 women), a traditional gendered division. However, deeper analysis and prolonged community engagement exposed the multiplicity and renegotiation of gender identities and livelihood roles, and the connection to declining fisheries. The relationship between gender and declining fisheries are considered in the following themes: community vulnerability; safety; education; and alternative livelihoods. The discussion shows that women experienced both emancipation and oppression in livelihood adaptation.

Community Vulnerability

Vulnerability in climate change is "the state of susceptibility to harm from exposure to stresses associated with environmental and social change and from the absence of capacity to adapt" (Adger 2006, p. 268). This includes the capacity to respond to external stress placed on livelihoods and wellbeing. Poverty is a vulnerability risk-factor, and in 2001, 63.3% of the Piura region was in poverty, of whom 21.4% were in extreme poverty (Herrera 2002). Small-scale artisanal fishers in Perú are vulnerable to climate change, as their adaptive capacity is low due to poverty and low education levels (Daw et al. 2009). Furthermore, women's vulnerability is magnified by both poverty and gender inequalities (Pettengell 2010), highlighted in Reyes (2002) study of the 1998 El Niño in the Piura region.

Community livelihood vulnerability was evident in Lobitos. Community members expressed their concerns about the future of fishing, and implications for families and young people. In the household survey, 38% of participants identified *falta de trabajo* [lack of work] as a community issue. "*Trabajo para todos*" [jobs for all] was a key policy platform in the 2010 local municipal election. In periods of severely low fish yields, I observed many men seeking alternative casual employment, several women undertaking poorly-paid domestic work, and some children not attending school due to lack of school uniforms and food.

Safety

In 2005, the World Health Organisation found that 51.2% of urban women and 69.0% of rural women in Perú had ever experienced physical or sexual violence in their relationship (Garcia-Moreno et al. 2005). The Peruvian 2010 Demographic Health Survey (INEI 2010) reports that in Piura, 39.8% of women had ever experienced either physical or sexual violence from their partner, and 14.6% in the preceding 12-months. Furthermore, 51.9% of these women had ever experienced physical or sexual violence when their partner was under the influence of drugs or alcohol, which is also a serious issue in Perú (Comisión Nacional Para el Desarrollo y Vida Sin Drogas 2003). Frequency of violence increased for women who were aged 40–44, divorced, separated, or widowed, with high school level of education, and poor.

In Lobitos, violence against women was a silenced issue, and was never mentioned in household interviews. However, when community trust with me strengthened, I was informed that several women in Lobitos experienced physical, sexual and financial abuse, which increased with partner's alcohol consumption that often increased with low fish yields.

> Rosanna, 42: Rosanna's parents were fishers, and her husband is a fisherman. She has four children. Her husband struggles to earn a sufficient income as a fisherman and drinks heavily. When he is inebriated he physically abuses Rosanna. He cannot find alternative employment, and Rosanna seeks casual domestic work to support the family.

Importantly, it cannot be assumed that a handful of anecdotes were representative of women in Lobitos. Furthermore, it was not completely clear whether declining fisheries was directly related to violence against women, and further research is certainly required. Nevertheless, women's personal stories indicated that the issues coexist and are potentially linked. Rosanna and Sofia's contrasting stories also challenge the dichotomous construction of *machismo* and *marianismo.*

> Sofia, 38: Sofia has never experienced family violence; however, she has knowledge of many other women who have. Following a family violence dispute in the community, Sofia visits the perpetrator and threatens to report him to the police if he continues to abuse. She explains that abuse is more common with alcohol.

Further research is required to explore the relationship between violence against women, alcohol and economic distress due to declining fish stocks. However, it is clear that safety concerns and associated disempowerment exist for some women in Lobitos.

Education

The lack of quality education for young people was raised by 53% of participating households in Lobitos. Community members believed that the traditional livelihood of fishing was not viable, and education was therefore very important. The community expressed their desire to learn useful skills and enhance their employability within the alternative livelihood of tourism. An English program was suggested by 70% of household survey participants, for children, young people, and adults to communicate with the tourism market. Furthermore, 43% of the participants supported WAVES' surfing program, which enabled understanding of the surf tourism culture. In response, WAVES developed education programs of surfing, tourism, small business, English, environment, surfboard repair, surf photography, and others as necessary. This process involved a renegotiation of traditional gender constructs. The majority of local people who were employed and trained to teach these programs were women, who gained positions of public leadership and developed tourism cultural capital through interacting with foreign volunteers. Additionally, several girls learned non-traditional skills for future work options, and boys considered alternative occupations to fishing. I also observed several local women commence part-time study to improve their employability. Enhanced education for women demonstrates emancipation from socialised gender barriers, and expose the multiplicity of feminine identities.

Maria, 12: Maria's father is a fisherman and her mother manages the family household and a small home-based shop. Maria participates in the WAVES education program and is learning to surf and repair surfboards, and studies English. Her parents want Maria to learn tourism-related skills because they recognise surf tourism as an alternative livelihood to fishing.

Alternative Livelihoods

The research data indicates that fishing does not provide sufficient work and income for families in Lobitos, and often becomes a fall-back profession when alternative work is not available. However, there are insufficient work opportunities in the community, and employment is generally casual and temporary (such as hotel construction and infrastructure projects). Many women challenged the *marianismo* construct and began working, a trend documented in the literature. Chant and Craske (2003) report that due to the decrease, informalisation and casualisation of employment in Latin America, men experience difficulties in maintaining their former privileged position and fulfilling their normative breadwinner role. Faced with their husband's unemployment, Perúvian women are likely to undertake income-generating activities (Ilahi 2001a); a process of "degendering" work (Fuller 2001).

In some circumstances, women's increased participation in the workforce was emancipatory. Three women who established small restaurants in their homes renegotiated their gendered roles to generate income, sometimes earning more money (and more consistently) than their fishermen husbands or sons. Power associated with the 'breadwinner' function shifted through a subtle subversion of labour, challenging the household's patriarchal foundation. In another example of gender renegotiation, two girls participated in WAVES' surfboard repair program, gaining skills to work in a traditionally masculine sector. Additionally, WAVES developed a pilot micro-credit program, providing training and loans to local women to initiate small businesses.

> Karina, 19: Karina participated in the WAVES microcredit program and developed a small business selling consumables to fishermen at the pier (a traditionally masculine space). She self-manages her business and earns a competitive local wage.

Although unreported and unobserved, the transition to alternative livelihoods could have been potentially dangerous for women, as previous literature suggests that Latina women who enter the workforce are at risk of violence. Flake (2005) found that employed women in Perú who have more education than their husbands and are the dominant decision-makers have increased likelihood of violence, which is further fuelled by alcohol. Flake hypothesises that men use violence to sustain their control over women, challenging the threat of women's employment to their masculinity.

Finally, livelihood adaptation to declining fisheries was limited by numerous factors in Lobitos. There were very few alternative employment opportunities, and a severe lack of water for alternative agriculture. Poor local infrastructure and limited community social and economic capacity constrained tourism business development. Tourism was also threatened by unpredictable El Niño events and potential damage to the wave and beach. Corruption by local authorities, pervasive nepotism and poor local infrastructure also affected individual and family options.

Conclusion

Declining fisheries significantly impacted women, girls, men, and boys in Lobitos. Men's inability to economically provide for their families, and women's increased participation in the public economy, undermined both *machismo* and *marianismo*. Within the themes of vulnerability, safety, education and alternative livelihoods, gender was diversified. In the process of gender renegotiation, women experienced both emancipation from traditional gender roles, and oppression through gender role reinforcement.

Firstly, the community desire to improve educational opportunities for children, including girls, implied a progressive community commitment to universal education as a pathway from poverty. This contemporary perspective contrasts with 1980s research that showed familial gender discrimination in decision-making regarding paying for schooling (Gertler and Glewwe 1992). National secondary education completion rates still, however, show a gendered difference, with 64.1% for women and 78.6% for men; and stark disparity between urban and rural women (65% versus 20.7%) (INEI 2010). Secondly, the emergence of alternative livelihoods in Lobitos may have increased empowerment and independence for some women. The activities of education, business development, and economic earning challenge the dichotomous gender paradigm. However, the risk of violence against women, indicated in previous research, suggests entrenched *machismo*. Thirdly, livelihood adaptation in Lobitos expose issues of vulnerability and safety for women, with risk of exploitation and violence due to power transitions. Finally, women's desire for structural change is evident, as expressed in the following quotation from a focus group: "Las madres están en casa lavando, cocinando y cuidando niños, y quieren otras cosas para hacer" [The mothers are in the home washing clothes, cooking and caring for the children, and want other things to do].

Sperling et al. (2008) explain that "strengthening the adaptive capacity to climatic shocks is vital to the sustainable development of the coastal and highland regions of Perú" (p. 65). The diversification of livelihoods due to declining fisheries has affected gender roles and relations in Lobitos. The community's experience of livelihood transition demonstrates the multiplicity of gendered identities, and shows that women can be both empowered and oppressed through climate change adaptation. Natalia's story provides a beacon of hope in an environment of gendered uncertainty.

> Natalia, 35: Natalia has a degree in business management, and is married to Juan, a fisherman and qualified plumber. They have five children. Unable to find work in their respective professions, the family relied on Juan's unreliable fishing income and struggled financially. After two decades of economic difficulty, Natalia engaged in professional up-skilling with WAVES. She now works full-time at WAVES, managing local staff and working with international volunteers. She and Juan co-manage childcare and household tasks, and she has also returned to university for further education, supported by a WAVES scholarship.

References

Adger WN (2006) Vulnerability. Glob Environ Change 16(3):268–281

Adger WN, Huq S, Brown K, Conway D, Hulme M (2003) Adaptation to climate change in the developing world. Prog Dev Stud 3(3):179–195

Alcalde MC (2010) The women in the violence: gender poverty and resistance in Perú. Vanderbilt University Press, Nashville

Alston M, Witney-Soanes K (2008) Social impacts of drought and declining water availability in the Murray-Darling Basin. Institute of Land, Water and Society, Charles Sturt University, Wagga Wagga

Angrosino MV (2008) Recontextualising observation: ethnography, pedagogy, and the prospects for a progressive political agenda. In: Denzin NK, Lincoln YS (eds) Collecting and interpreting qualitative materials. Sage, Thousand Oaks, pp 161–184

Baas S, Ramasamy S (2008) Community based adaptation in action: a case study from Bangladesh. Food and Agriculture Organization, Rome

Badjeck M-C, Allison EH, Halls AS, Dulvy NK (2010) Impacts of climate variability and change on fishery-based livelihood. Mar Policy 34(3):375–383

Basham R (1976) Machismo. Front J Women Stud 1(2):126–143

Beattie P (2002) Beyond machismos: recent examinations of masculinities in Latin America. Men and Mascul 4(3):303–308
Bertrand A, Segura M, Gutierrez M, Vasquez L (2004) From small-scale habitat loopholes to decadal cycles: a habitat-based hypothesis explaining fluctuation in pelagic fish populations off Perú. Fish Fisheries 5:296–316
Brandth B, Haugen MS (2010) Doing farm tourism: the intertwining practices of gender and work. Signs 35(2):425–446
Chant S, Craske N (2003) Gender in Latin America. Rutgers University Press, New Brunswick
Comisión Nacional Para el Desarrollo y Vida Sin Drogas (2003) II Encuesta Nacional Sobre Prevencion Y Consumo De Drogas 2002. Comisión Nacional Para el Desarrollo y Vida Sin Drogas, Lima
Corbin JM, Strauss L (2008) Basics of qualitative research: techniques and procedures for developing grounded theory. Sage, Thousand Oaks. http://www.google.com.au/search?tbo=p&tbm=bks&q=inauthor:%22Anselm+L.+Strauss%22
Daw T, Adger WN, Brown K, Badjeck M-C (2009) Climate change and capture fisheries: potential impacts, adaptation and mitigation. In: Cochrane K, De Young C, Soto D, Bahri T (eds) Climate change implications for fisheries and aquaculture: overview of current scientific knowledge: FAO Fisheries and Aquaculture Technical Paper, No. 530. FAO, Rome, pp 107–150
Ehlers TB (1991) Debunking marianismo – economic vulnerability and survival strategies among Guatemalan wives. Ethnology 30(1):1–16
FAO (2010) National fisheries sector overview. FAO, Rome
Flake DF (2005) Individual, family and community risk markers for domestic violence in Perú. Violence Against Women 11:353–373
Fuller N (2001) The social construction of gender identity among Perúvian Men. Men Masculinities 3(3):316–331
Garcia-Moreno C, Jansen AFM, Ellsberg M, Heise L, Watts C (2005) WHO multi-country study on women's health and domestic violence against women, world health organisation: initial results on prevalence, health outcomes and women's responses. World Health Organisation Press, Geneva
Gertler P, Glewwe P (1992) The willingness to pay for education for daughters in contrast to sons: evidence from rural Perú. World Bank Econ Rev 6(1):171–188
Hays-Mitchell M (1995) Voices and visions from the streets: gender interests and political participation among women informal traders in Latin America. Environ Plan Soc Space 13(4):445–469
Herrera J (2002) La pobreza en el Perú en 2001: Una visión departmental. INEI, Lima
Ibarra AA, Reid C, Thorpe A (2000) The political economy of marine fisheries development in Perú, Chile and Mexico. J Lat Am Stud 32:503–527
Ilahi N (2001a) Children's work and schooling: does gender matter? Evidence from the Perú LSMS panel data. The World Bank, Washington, DC
Ilahi N (2001b) Gender and the allocation of adult time: evidence from the Perú LSMS panel data. The World Bank, Washington, DC
INEI (2007) Censos nacionales 2007.http://censos.inei.gob.pe/censos2007, http://censos.inei.gob.pe/censos2007. Retrieved 14 Aug 2011
INEI (2010) Encuesta Demografica y de Salud Familiar 2010: Nacional y Departmental. Instituto Nacional de Estadistica e Informatica, Lima
IPCC (2007) Fourth assessment report: climate change 2007, working group II: impacts, adaption and vulnerability. IPCC, Geneva
Jorgensen DL (1989) Participant observation: a methodology for human studies. Sage, Newbury Park
Lambrou Y, Grazia P (2006) Gender: the missing component of the response to climate change. Food and Agricultural Organization, Rome
Marshall C, Rossman GB (2011) Designing qualitative research. Sage, Thousand Oaks
Mirande A (1997) Hombres y machos: masculinity and Latino culture. Westview Press, Boulder

Mitchell T, Tanner T, Lussier K (2007) We know what we need: South Asian women speak out on climate change adaptation. Institute of Development Studies and ActionAid, Johannesburg
Olivas-Lujan MR, Monserrat SI, Ruiz-Guiterrez JA, Greenwood RA, Gomez SM, Murphy EF, Bastos F. Santos NM (2008) Values and attitudes towards women in Argentina, Brazil, Colombia and Mexico. Empl Relat 31(3):227–244
Padilla B (2004) Grassroots participation and feminist gender identities: a case study of women from the popular sector in metropolitan Lima, Perú. J Int Women's Stud 6(1):93–113
Pauly R, Christensen V, Dalsgaard J, Froese R, Torres F (1998) Fishing down marine food webs. Science 279(5352):860–863
Peña M (1995) Feminist Christian women in Latin America: other voices, other visions. J Fem Stud Relig 11(1):81–94
Pettengell C (2010) Climate change adaptation: enabling people living in poverty to adapt. Oxfam Great Britain, Oxford
Reyes RR (2002) Gendering responses to El Niño in rural Perú. Gend Dev 10(2):60–69
Safa H (1990) Women's social movements in Latin America. Gend Soc 4(3):354–369
Sperling F, Vilidivia C, Quiroz R, Vilidivia R, Angulo L, Seimon A, Noble I (2008) Transitioning to climate resilient development: perspectives from communities in Perú. World Bank, Washington, DC
Stevens EP (1973a) Marianismo: the other face of machismo in Latin America. In: Pescatello A (ed) Female and male in Latin America. University of Pittsburgh Press, Pittsburgh, pp 88–101
Stevens EP (1973b) The prospect for a women's liberation movement in Latin America. J Marriage Fam 35:313–331
Thomas D, Twyman C (2005) Equity and justice in climate change adaptation amongst natural-resource-dependent societies. Glob Environ Change 15:115–124
UNDP (2010) Human development report 2010. The wealth of nations: pathways to human development. United Nations, New York
WAVES for Development (2012) Who is WAVES?http://www.wavesfordevelopment.org/who-is-waves, http://www.wavesfordevelopment.org/who-is-waves. Retrieved 20 Feb 2012
Young KR, Leon B (2010) Natural hazards in Perú: causation and vulnerability. Dev Earth Surf Process 13:165–180

Chapter 19
Climate Change: A Himalayan Perspective 'Local Knowledge – The Way Forward'

Reetu Sogani

Abstract This chapter reflects the insights of people in Bheerapani and surrounding areas with whom discussions and community based interventions were carried out. This work, carried out over 9 years, was undertaken with minimal funding and would not have been possible without the active participation of Bhavan Singh, Mohan and Vimala Karki of Community Awareness Centre and the author herself as the co ordinator.

Keywords Climate change • Himalaya • Agricultural practice • Traditional farming • Changing weather patterns • Adaptation strategies

Introduction

Life in mountain ecosystems all over the world has certain commonalities such as fragility, inaccessibility, marginality and diversity and these have a huge impact on the everyday life of people living in such regions. Survival needs in these difficult and unique biophysical conditions have resulted in the evolution of 'human adaptation mechanisms' or knowledge systems. These complex and proven systems and practices have emerged through the process of continuous observations, experiences, experiments and adaptation by local people.

The Himalayas are not only the most magnificent of all mountain systems, but being young, ecologically fragile and densely populated, they happen to be one of the most threatened ecosystems in the world. Uttarakhand, in Central Himalaya, has a rich repository of traditional knowledge which forms a crucial component in the survival strategies of people living in this Himalayan environment. This chapter

R. Sogani, Ph.D. (✉)
Chief Co ordinator (Community Awareness Centre). Independent Development Practitioner, Uttarakhand, India
e-mail: reetusogani@yahoo.com

M. Alston and K. Whittenbury (eds.), *Research, Action and Policy: Addressing the Gendered Impacts of Climate Change*, DOI 10.1007/978-94-007-5518-5_19,

reflects the insights of people in this region rich with the tremendous variety of different grains, crops and many unique agricultural practices.

Agriculture is the main source of livelihood (at least 70–75% of the population depends on it) here and the majority of the farmers are small or marginal with average land holding size not more than an hectare. With only 12–13% of the land being irrigated, agriculture is primarily rain fed and is dependent on rainfall.

Climate Change in Himalaya

Being an elevated region, Himalaya is certainly more sensitive to and more affected by climate change. For academic reasons, mountain areas provide conducive conditions for studying changes in temperature and weather patterns along with adaptations being made. Due to high elevation, mountainous regions act an early warning system to climate change. Moreover, there is evidence to prove that Himalayas are warming 5–6 times more than the global average (Singh n.d.) an alarming trend as they are the world's third largest reservoir of water after the polar caps. Being connected to nine river basins, they are the life line to an unimaginable population including 500 million people living in Gangetic river basin alone.

Central Role of Women in Himalaya

Women are the backbone of mountain life in this part of the Himalaya. Despite the favourable sex ratio and comparatively higher literacy levels, the status of women in the state of Uttarakhand is quite dismal. Women bear an inordinately larger portion of the domestic and economic workload (60–70%). Besides household chores, most activities relating to agriculture, natural resource management and animal husbandry are carried out by women. "80–90% of the agriculture work is being done by women" (Kaushik 2004).

Being closely connected to natural resources, it is not surprising that women possess invaluable knowledge. This enables women to adapt to difficult situations and to eke out a decent existence even in uncertain conditions. Despite being the mainstay of the economy and the holders and practitioners of local knowledge, women have very little say in the decision-making process relating to agriculture, water and forests and this remains a male domain.

In earlier times, women made most of the important decisions relating to the choice of crop selection, land use, harvesting schedule and agricultural inputs. This gave women status and autonomy. However, new agricultural systems influenced by market forces, relying heavily on chemical inputs, are gradually reducing women to unskilled labourers. Government policies and programmes are not targeted towards women, as they are not considered farmers, despite significant contribution to the agricultural economy.

Changes in Agricultural Practices

According to local informants, cash crop farming introduced in the 1970s through the green revolution has gradually replaced local resource based sustainable ways of farming. The current system is dependent on markets both for agricultural inputs as well as the sale of produce.

This system, apart from being loaded with chemicals, is highly water intensive, removes friendly pests and insects from the soil, reduces soil fertility and involves high input costs. Not surprisingly, many of the farmers practising this kind of agriculture are now under heavy debt. Moreover, it has added more work hours to already burdened women. While earlier, traditional crops such as millet provided a good source of residue for fields and animal feed, now women have no option but to walk miles for fodder and biomass. This has also brought about forest degradation. Women feel that this system has made people food and nutritionally insecure. Women's health is further worsened with about 45.6% of the women in Uttarakhand suffering from anaemia (National Commission on Population, Government of India 2000).

Traditional Ways of Farming

Local people feel that traditional crops and ways of farming such as mixed cropping, rotational cropping and other methods, ensured better nutritional balance and more food security in terms of variety and quantity as well as an insurance against weather variability and pest or insect attack. Farm yard manure not just fulfills the macronutrient needs, but also the micronutrient requirements of crops and soils. According to women "chemical fertilisers burn the soil and we have to use cow dung manure alongside to prevent this". This also helps retain moisture in the soil. Most traditional crops, being tough, are also less dependent on water and can withstand long dry spells. Some traditional crops, especially millet, have long tough roots that bind the soil together, thereby preventing soil erosion.

Local Wisdom

According to Hemanti, a dalit (socially discriminated class) leader, "Mixed cropping ensured the presence of at least some food all the time, if one grain would get over today, another would be ready for harvest tomorrow. Children would always have something to eat" (*translation*). This kind of farming also ensured cultivation in small fields and the presence of grains, lentils, vegetables even in poor houses in the hills.

Traditional systems of farming required even less labour for a good bumper harvest. "This kind of farming would require less labour", was noted by many women in the hills. This is also evident from the statistics – which indicate that paddy cultivation involves most time (1,155 h), followed by wheat (1,010 h) and vegetables (920 h). In comparison coarse millets are much less demanding with Madira (barnyard millet) and Mandua (finger millet) taking up just 564 h in a year (Sogani 1998).

Thus it appears that cultivation of coarse crops is most suited to the hill environment, is nutritious, less labour intensive and less time consuming. This is contrary to the popular myth that traditional farming is more demanding in terms of labour and time.

Women reveal that local knowledge based agriculture has several nutritional advantages as well. It provides balanced nutrition and high levels of minerals such as calcium and iron through traditional crops such as Mandua. "We used to eat Mandua and that is reason our generation is still strong while the new generation feeding on food items from the market, is physically much weaker"says Parvati, a 60-year-old woman, who also happens to be the first headperson of a formal forest committee in the Bheerapani area.

In short, this kind of farming, not just ensured balanced food on the plate, it also ensured environmental sustainability and economic viability. This is important and relevant in the wake of the changing global scenario.

Changing Weather Patterns

Higher frequency of extreme weather events such as droughts and floods, as well as greater unpredictability and variability in the seasons and rainfall is increasingly being felt the world over. People in Himalaya are more vulnerable to changes in weather patterns and their impacts. In fact, they strongly believe, based on their experience that what appears to be climate change induced might be more of an outcome of environmentally unsustainable and economically unviable development policy in agriculture, water, forestry and other sectors including infrastructure and power.

According to a study carried out by the Wadia Institute of Geology, Dehradun, in Ukhimath region (high altitude region in Garhwal), landslides and soil erosion have occurred more in those areas where chemical intensive farming and cash crop cultivation is being practiced (Jardhari, 2010, Farmer and activist in Uttarakhand, Beej Bachao Andolan, India, personal communication).

In the mountain ecosystem, with agriculture being based on natural resources and climate sensitivity, farming has taken a serious setback. According to Parvati (local informant), "earlier weather related changes would happen slowly but now we have been facing ups and downs in weather so fast that it becomes difficult sometimes to even cope with them". This is being experienced by most of the rural community, including Anandi, who says, "we have been hearing and experiencing changes in weather for a long time now with rainfall declining, snow fall receding, and production decreasing".

It is obvious that many of the local women farmers cannot relate weather variability to a global phenomenon of increasing carbon emissions and global warming. Nonetheless, they have a rich storage of information on the impact it is leaving on their agriculture, natural resources and livelihoods.

Changes Experienced Locally

According to local people, there used to be at least four to five snowfalls from November to February each winter. However, these snowfalls have reduced substantially over recent years, with some people experiencing as little as just one snowfall during the entire winter. With declining snow fall and winter precipitation, there is less moisture for growing crops. According to Munni, a 55-year-old woman, with a good apple orchard, "we lost some very good varieties of apples, with good taste and production a few years ago. Low temperature accompanied with snow and ice was instrumental in having a good harvest and taste". Similar experiences are occurring in other areas as winter cold has shifted by at least a month and so has the accompanying snowfall. Weather records match people's perception of summers becoming hotter than usual and winters becoming warmer. Highest maximum temperatures over the last 10 years were significantly higher in the month of January in Nainital and Dehradun districts. Lowest minimum temperatures are also rising (Sogani 2011).

Local people are worried by their observations that the size of glaciers is reducing each year. They believe their survival is connected to the vast glaciers and note that the Himalayas are not looking 'white' anymore, "We can see large patches of naked mountains now, and it has been increasing every year. Our life is connected with them. Without them, we will die away gradually".

While annual precipitation has not changed greatly, its timing, frequency, duration and intensity has certainly undergone a change. Besides being highly erratic, incidents of extreme weather events and shifting seasons are becoming more frequent. According to elderly women, this kind of rainfall increases runoff resulting in reduced groundwater recharge while 'slow and steady' precipitation is beneficial.

Not surprisingly, the availability of water in natural water sources such as 'Naulas' (natural ground water springs) and 'Gadheras' (streams flowing in valleys) has reduced considerably. During 2010, the hilly region saw unprecedented rainfall for almost a week. Despite this, water levels in the some water sources were quite low 1 month later. In Uttarakhand, almost 30% of water sources have dried up, while the remaining 70% have severely weakened over recent years (Jardhari, 2010, Farmer and activist in Uttarakhand, Beej Bachao Andolan, India, personal communication).

Heavy rainfall, previously unknown at high altitude, has become more frequent causing flash floods that wash away homes, fields, trees and livestock. Extreme weather events have been experienced more frequently over recent years. In 2010 the region faced a long dry spell followed by heavy rains. Heavy floods hit the area in 1993, 2003 and then in 2011 causing widespread damage to life and property.

Other changes are also being experienced. Spring is coming earlier. Sometimes it is not the 'real' spring as it can be followed with a blast of cold weather. Hailstorms, not previously a common occurrence, have become more frequent and untimely. According to women, small sized hailstones were good for cultivation. Whereas now, with the size of hail stones sometimes larger than a cricket ball, destruction of crops especially fruits is substantial.

Impact of Changes

Increasing temperatures have resulted in early maturity of crops and changed timings with respect to budding of flowers, leaf fall and harvesting of grains. As a result people living at altitudes between 1,500 and 2,000 m now sow their crops one to one and a half months earlier. Likewise, harvesting has also been brought forward by almost 2 months. These changes have also led to intensification of land use. Unlike in the times of traditional agriculture when two crops were harvested annually, farmers are now taking three crops, which is primarily to maximize economic returns from farming. This practice has not come without a cost. According to some women, it has further disturbed the fertility of soil as "we are not giving any opportunity to the soil to breathe and regain its strength".

Some higher altitude (1,500–2,200 m) varieties such as 'Aadu' (peach), 'Khubani' (apricot) now have a lower yield for the want of low temperatures. Conversely, production of tropical fruits like banana, guava, lychee, papaya and even mangoes has increased in the valleys. Production has been adversely affected by off-season flowering. This has resulted in flowers becoming shriveled and weak fruit affecting both quality and quantity of yields. This has affected the traditional livelihood avenues of hill people as markets are depleted of quality local varieties.

Changes have noticeably affected the natural flora and fauna as well. 'Buransh' (Rhododendron) used to flower in March-April but one can see blooms even in the months of December and January. Likewise, other forest fruits such as 'Kafal' (a local fruit forest tree) are blooming well before their scheduled summer time. There has been a noticeable vertical migration of species. 'Gait' (a tuberous plant found in mixed forests), normally found at a height of 1,500 m, can now be found at an altitude of 2,200 m. Bird species of the plains like the Common Mynah are now seen regularly at high altitudes and it is not uncommon to hear the songs of the Koel (Indian Cuckoo) at high altitudes.

Observations of local women in regular contact with forests and other natural resources show that the composition of forests has also undergone a change with loss of genetic variety in the forest. This has occurred because some traditionally found species are unable to grow in the forest as the amount of moisture they need is not timely or is missing completely.

People have also experienced an increase in pests and crop diseases. They believe that this is the result of excessive use of pesticides and urea and is not merely an outcome of the warmer climate. Nevertheless, the increase in temperatures along with other accompanying factors, has led to mosquitoes, pests and houseflies from the foothills now surviving in the higher hills. The presence of these insects suggests the possible spread of vector-borne diseases, such as malaria and dengue fever to areas where cooler temperatures previously protected people from these threats. Citrus fruits are becoming affected by fungal infection. Likewise, peaches and apricots at lower altitudes are now becoming diseased.

With snowfall and rain being two important sources of water in the hills, reduced snowfall impacts water recharge, thereby reducing water levels in streams and the ground water. Erratic rainfall has resulted in crop failure and this is particularly evident in chemical intensive agriculture which is more water intensive.

Local informants report that encroachment of forest area for agriculture has had serious impacts. They suggest that this was probably one of the causal factors for the washing away of huge land portions in recent floods. Faulty road construction is another factor making the region vulnerable to increasing landslides and soil erosion. This has also adversely affected the land productivity of the area.

Cultural upheaval is another impact of these climatic changes. Earlier agricultural calendars used to coincide with cultural festivals such as Phuldeyi (the festival of flowers). However, rhododendrons and other flowers now bloom earlier, in December and January, while the festival of flowers occurs later, in mid-March. Similarly, on the last day of the Navratri (festival of nine nights) during the summer season, wheat stalks, ready for harvest, are roasted and offered to gods. Nowadays, by the time last day of Navratri is celebrated, crops have often already been harvested. With crop timings going haywire, the cultural significance of these festivals is in danger with possible adverse impacts on the social fabric of the region.

Repeated crop failures have also resulted in the depletion of seeds in villages with crop production affected in the long term. According to women and others, one crucial negative outcome of the green revolution was the gradual erosion of seeds and other forms of biodiversity.

Climate changes and the policies and programmes of development agencies, including the government, have gradually led to changes in the food habits of people. Consumption of milk and its derived products has decreased while tea, alcohol and soft drinks have gained in popularity in recent years. Home use of processed and packaged food is increasing while locally grown and produced food is decreasingly used in homes. Even rural homes are not immune from this phenomenon with traditional millets, local vegetables and pulses mostly being replaced with rice, wheat and other marketed groceries.

Forest degradation, reducing common property resources as well as replacement of traditional crops with cash crops has not only reduced the availability of fodder and grass but made women's tasks more difficult and laborious. It has likely resulted in families becoming more food and nutritionally insecure and people becoming more vulnerable to disease and hunger, especially women.

Adaptation Strategies and Their Impact

After experiencing these fluctuations in weather patterns for many years, people in the Himalayas are gradually making a shift in terms of their cropping patterns. Some of the villages have started replacing food grain production such as wheat with vegetable production such as green chillies, peas, 'dhaniya' (coriander) and also spices such as turmeric and ginger that require warmer climatic conditions. More acreage is now assigned to fruit production in the valleys. People report that they are practising this pattern of crop diversification because it involves less investment in terms of time, labour and money. Although income producing, this strategy has its drawbacks. With families shifting more to cash crops requiring increased use of fertilisers and pesticides to maintain the level of production as well as their increasing dependence on market, there is a greater likelihood of their food and nutritional security significantly decreasing. As noted by Anandi – "though this is getting us money it has a dark side as well, as we eventually get dependent on the market to buy food and inputs for agriculture".

Forest degradation has also affected lives of farmers with more frequent migration of animals into villages and farm fields. "We are now forced to grow crops which are not being liked by wild boars such as ginger and turmeric". Wild animal attacks have also brought about a change in cropping pattern which is not always socially sustainable.

With food prices soaring and local food produce not lasting for more than four months a year, ensuring food availability and nutritional security at household level is becoming a huge challenge for women. This has brought about an increasing dependence on government schemes such as public distribution systems and employment guarantee programmes. Although this brings in money to buy food from the market, it leads to consumption of inferior quality foodstuffs.

With agriculture losing its economic viability due to a variety of factors, including reduced labour availability, people, especially men, are being forced to leave farming and migrate to cities to take up wage labour. In some cases this results in leaving land fallow and reducing the amount of land under agriculture. This migration leaves women behind with even more responsibilities relating to household chores and farming. This usually results in an increased workload for women and all the added social and emotional stress that it creates, for both those who migrate and those who are left behind. Moreover, the condition of women has worsened with no ownership of land and almost complete absence of access and control over productive resources. To ensure better food availability, women are also undertaking wage labour but, unlike men, they prefer to take up jobs within or close to their villages.

With less land under cultivation, food related crises are likely to increase in the future. People who are still practising agriculture are increasingly taking loans and credit to survive. It is now common for farmers to be in heavy debt.

People have been hard hit during extremes of precipitation wherever cash crops have made inroads as hybrid varieties are affected more (than traditional crops) during both drought and flood. Conversely, traditional millets and local varieties of rice survived drought and flood more successfully, although they also incurred considerable damage in terms of crop and seed loss.

According to one progressive farmer, "Millets survived the drought conditions more successfully since their water requirement is quite low. In fact, during the drought period, their production reached a new high while production was affected a bit when it rained heavily". A similar experience was shared by another farmer, "Last season when there was a long dry spell, we had a good crop of Mandua while all the other crops failed".

Case Study of Bheerapani

A small rural based organization, Community Awareness Centre (CAC), is working in Central Himalaya, to raise awareness around traditional knowledge issues, improve the opportunities and options of livelihoods based on these knowledge systems and build necessary conditions which will promote local people's inclusion in the development process. Situated at an altitude of 2,200 m, in a small village called Bheerapani, CAC is working to use people's knowledge to promote the empowerment of marginalized groups as well as ensure the protection and stability of fragile ecosystems.

Since the advent of chemical intensive farming, women have been participating less in decision making processes in the family as well as at the community level. Strategies adopted to cope with the changing situations have not been gender sensitive and empowering. This is usually the situation where decision makers are mostly men, with women not usually being part of any economic decision making.

However in Bheerapani and other places where farmers, both men and women, are still rooted to their environment, women are in the forefront and are participating in formal decision making processes. They have been drawing on their rich body of knowledge to adjust to the changing weather situation using adaptive and mitigating strategies.

Nine years ago women came together in Bheerapani and constituted strong community based organisations to address issues in agriculture, forestry, water, health and even social issues and contributed to the development of the village. Participatory exercises on issues including women's decision making relative to responsibilities encouraged them to assess, analyse and develop a community based plan replete with responsibilities and a time line. Although people had been experiencing changes in weather patterns for quite some time, they did not fully appreciate the need to act specifically to address climate change. However, their use of strategies to bring about sustainability in agriculture and forest management enabled them to develop better resilience to climate change and its impacts.

Awareness generation workshops, exposure and expedition trips, capacity building exercises to produce improved composting techniques and local resource based growth promoters encouraged women to revert to biodiversity based farming.

Women and some progressive male farmers started replacing cash crops, particularly vegetables, with grains such as wheat and millet. Retrieval of traditional seeds and varieties helped them set up a traditional seed bank. Women being holders and practitioners of traditional knowledge systems and practices, easily realized that biodiversity based ecological farming is the sustainable way forward. However, with land being legally owned by men, who were skeptical, it was not an easy path. Patient discussions, workshops and informal interactions were carried out regularly with men. With people having lost confidence and belief in their own systems and practices, eventual adoption and conversion to sustainable practices took place very slowly initially but gradually gained ground with better results. Besides reducing their dependence on the market, the possibility of better access to nutritious foods became apparent.

To reduce runoff, prevent soil erosion and improve ground water recharge, people have constructed check dams using government schemes. Biogas units have also been installed in several households. Besides providing a fuel which is economic and efficient, which also reduces dependence on forests for fuel wood, its slurry is excellent compost for plants.

Due to women's active participation in informal and then in formal forest committee meetings as headpersons and members, mixed forests have improved considerably with respect to their composition and density. Dependence on forest resources has decreased due to availability of residue from traditional crops. With traditional crops gaining popularity, women have also experienced a decrease in their workload. Moreover the improved condition of the forest has reduced everyday distance travelled by women to collect biomass. Animal health has also improved due to better quality feed from traditional crops and the improvements in forest conditions.

Experiences and incidents from Bheerapani, as well as other areas of the hills, illustrate that ecological farming and practices based on local resources and local wisdom and knowledge have helped women and the rural community face adversity associated with climate variability with more resilience.

Conclusion

The Himalayas are one of the world's most ecologically fragile ecosystems and are of crucial importance to the entire world. Climate change is very apparent in the Himalayas and has direct impacts on the lives of people living within and beyond the region. Most development policies in recent years have been resource intensive and environmentally unsustainable, which has undoubtedly aggravated the impact of climate change. Many bio-diverse farming systems are also effective climate change mitigation strategies as they mitigate carbon emissions and can substantially mitigate global warming impacts through carbon sequestration.

Women are the mainstay of the agrarian rural economy in the Himalayas but they have been relegated to the background and women's knowledge, experience and potential contribution in developing effective coping strategies for climate change is yet to be harnessed. Women's work load has increased and intensified due to degraded common property resources, reduced water levels, increasing cropping cycles and a decline in interdependence and co-operation amongst local people in farming, grass cutting, weeding and other tedious activities.

Global modes of production, consumption and trade have been largely responsible for undermining local resources and knowledge systems over recent decades. This has resulted in massive environmental destruction and has disturbed local autonomous and sustainable modes of livelihood activities. These changes have exacerbated the impact of climate change and have made people, especially women, more vulnerable to food insecurity, disease and hunger. With a background of chronic high levels of malnutrition and anemia amongst women and girls, these problems have become more acute.

Local people's recent experiences have strengthened their belief in biodiversity based ecological farming and local knowledge based ways to use and manage forest reserves and other natural resources. Being environmentally, economically and socially sustainable, they prove to be an effective strategy in response to the impact of climatic variability and climate changes. In short 'Local knowledge is the way ahead'.

Governments and development agencies may differentiate between climate change adaptation, poverty alleviation, food and nutritional security. However, eventual goals are wellbeing of humans, sustainability of the environment and protection of the earth. Therefore a trans disciplinary and holistic approach is needed in the consideration of climate change.

Perceptions, experiences, insights and strategies adopted by women and indigenous people all over the world need to be given due importance and recognition by the world community in climate change discussions. Mitigation and adaptation strategies based on their wisdom and time tested knowledge, can help develop a better understanding of changes, impacts and coping mechanisms.

References

Kaushik S (2004) Situational analysis of women and girls in Uttaranchal. National Commission for Women, New Delhi

National Commission on Population, Government of India (2000) www.populationcommission.nic.in.ut.htm

Singe SP (n.d.) Climate change in relation to the Himalayas www.climate-leaders.org/wp-content//climatechange-spsingh.pdf

Sogani R (1998) Farmer without her Plough. Paper presented at third congress on traditional sciences and technologies of India, Varanasi, November. (FAO of the UN regional office for Asia and the Pacific). ftp://ftp.faoorg/sd/sdw/sdww/ind/pdf

Sogani R (2011) Documentation of climate change perceptions and adaptation practices in Uttarakhand, Northern India. Beej Bachao Andolan-Save Seed Campaign, PANAP, Malaysia

Chapter 20
Gender and Climate Change: Implications for Responding to the Needs of Those Affected by Natural Disasters and Other Severe Weather Events

Desley Hargreaves

Abstract This brief paper is a practice based paper and discusses the experience of social workers in the Department of Human Services (DHS) who have worked with people affected by natural disasters and other severe weather events. Over the past 10 years Australia has experienced a range of these events and DHS social workers have been involved in developing innovative responses to meet the needs of affected individuals, families and communities. The paper highlights the way in which responses are tailored to take account of gender differences in help seeking behaviours. It also briefly discusses the conscious decision making around gender balance in the teams who provide assistance to affected people including responses to offshore events affecting Australians.

Keywords Gender • Climate change • Natural disaster • Severe weather events • Responses

Introduction

Over the last 10 years Australia has experienced a range of natural disasters and severe weather events which have impacted individuals, families and communities. Australia is a land of great contrasts from rich flood plains and golden beaches to cattle country and sandy deserts. Drought has been a constant feature in Australia's climate history. However the impact of climate change and the effects of changing weather patterns across the Pacific Ocean (El Nino and La Nina) have seen the experience and duration of drought increase in severity. Severe weather events

D. Hargreaves (✉)
Social Work Services Branch, Commonwealth Department of Human Services, Canberra, ACT, Australia
e-mail: desley.hargreaves@humanservices.gov.au

M. Alston and K. Whittenbury (eds.), *Research, Action and Policy: Addressing the Gendered Impacts of Climate Change*, DOI 10.1007/978-94-007-5518-5_20,

resulting in flooding and damage to housing and infrastructure due to high winds have often impacted drought affected areas. These effects have been compounded in some areas by green drought, mice and locust plagues. The summer of 2011 in Australia was a classic example of that.

The Challenges

As workers responding to people and communities impacted by these events the challenge becomes one of relevance, helping to create meaning and hope in what is often experienced as a hopeless situation. Our experience has been that men and women respond differently to the disaster event and demonstrate different help seeking behaviours. We have also observed some differences in the way people from rural and urban areas respond to such events.

In our experience as a simplistic generalisation, men tend to focus on practical, physical matters such as rebuilding fences, searching for or feeding livestock, rebuilding sheds, ensuring a supply of food for the animals, on occasion shooting them to avoid further suffering; searching for bodies of loved ones, cleaning up. Likewise, in our experience women have generally tended to support their husband or partner but focus on the emotional wellbeing of their children, husband, other family members and neighbours. Self care has not been a strong suit for either gender. Men have tended to withdraw, internalise their feelings and not engage in help seeking nor respond to those wishing to provide support. Women often bear silent witness to this while trying to stay strong for their children and their partners. Lack of income can at times increase the isolation experienced by the whole family because basic social activities are simply unaffordable.

The challenge for service delivery is how to respond appropriately often within a limited timeframe, with maximum impact, building on strengths and promoting resilience.

In their review of the empirical evidence of the five essential elements of immediate and mid-term mass trauma intervention Hobfall and colleagues identify the Promotion of Connectedness as of central importance (Hobfall et al. 2007). I believe that this has a broader application.

DHS staff who have been working in drought affected communities talk about the challenge of social connection when year after year people have been impacted not only by the drought itself, but by the impact on economic viability of the farm, the need for one or both partners to work off the farm – this in itself challenges the traditional gender roles. There is also a declining income and sense of despair and hopelessness which often leads families to disconnect from their social support networks. In some drought and bushfire affected communities many leave and those who choose to stay can find themselves in a disintegrating community. These circumstances can often lead to people becoming quite cynical about any offers of support.

Downey (2010) notes that farm succession planning – a predominantly male oriented arrangement- has been impacted significantly by climate change. When succession and exit options become essential during times of dryness this can become an additional stressor. The farmer is reluctant to pass on a non-viable business and fears a loss of identity. Often the farmer's wife fears for her husband or partner. Depression is a common experience and suicide among farmers in these circumstances is not uncommon. In these situations DHS has used a combination of a Rural Service Officer and a social worker to work with both parties to address their financial entitlement and emotional needs. Linkages to existing programs such as Mates helping Mates, and Men's Sheds have also been important in making appropriate and responsive referrals.

Pollard and Oxenham (2010), two DHS social workers, have written about their involvement in the Family Support Drought Response Team project based in Mt Isa. Their area covered 611,908 km^2 (three times the size of the State of Victoria) with a population of just 73,800. They referred to comment from the Expert Social Panel on the Human Response to Drought: "People in drought communities were exhausted and too busy surviving to engage in social activities" as being particularly relevant to their experience.

These social workers engaged with existing networks such as the Department of Primary Industries, AGFORCE, North West Mental Health services and other services that were already connecting with men on drought affected properties. They were able to use strategically the DHS Drought Bus – a mobile office – and attend planned community events. Partnering with the Royal Flying Doctor Service and the School of the Air, they were able to connect with a twice yearly face to face meeting with home tutors. Pollard and Oxenham (2010) took a gendered approach working with rural women at these events, using group activities and workshops to maintain connection. They used these women's networks to reach out to men and other family members, particularly with the provision of information applicable to the general population including support contact information. On air support workshops, email and other technology also enabled them to connect in innovative ways with rural families.

Another social worker Lisa Roberts in the Drought Response Team in Central Queensland developed a novel way of attracting women's interest through her attendance at Agricultural shows, Agfests and other rural events which brought men, women and families together. She developed a Resilience tree (Roberts 2010) and encouraged women to write on specially made leaves, strategies that they used to cope with their difficult situation, to share with others. These leaves were attached to the Resilience tree. Lisa reported that wherever she used that strategy, women would come back and add a few more ideas, helping them to identify the strengths they had to build on. She used this idea effectively in workshops with rural women also. The idea was a non-threatening way for her to engage with women in the first instance and then use that opportunity to talk further with individuals or invite them to attend the workshops.

In January 2011 in the small Victorian town of Rochester flood waters caused significant damage. This was the last straw for many in the community which had also suffered from years of drought and also a locust plague. The rural social worker Emma who had been working with that community for some time on the drought issues, adapted a program for women called The Next Step – a 4-week strengths based program. Both day and night sessions were offered to accommodate availability of potential participants.

It involved storytelling, identifying strengths, coping with change, grief and loss and self care. It was organised as a partnership with a range of local agencies. Feedback from one participant indicated that it was the first time she felt heard by someone who cared and that it had given her permission to engage in self care, not just care for significant others (Department of Human Services 2011).

Engaging both men and women can be challenging and requires innovative strategies. In far western New South Wales social worker Helen coordinated a Chainsaws and Chocolates event after a discussion with the local school principal. It was linked to a 2-day event – the Home Tutor Conference. This involved parents, teachers and the broader community. Several community agencies became involved. The Royal Flying Doctor Service provided health checks and launched their mental health packs; financial management advisors, legal services, health and wellbeing promotions; employment service providers, local service agencies and pastoral care services were all available to everyone during that period. It helped to normalise the offers of assistance and to reach out to both men and women (Department of Human Services 2011).

The experience of DHS social workers has shown that is important to connect with people where they are. They won't necessarily come asking for help – so farm visits, farm gate events, agricultural shows and other organised community events – where help seeking and response can be normalised – are important responses used by our department.

Finally I want to provide a brief comment on gender balance in teams of workers responding to both offshore and onshore events. Since 2003 following the Bali bombings in which 88 Australians were killed the Australian government has included DHS social workers in its crisis response teams on many occasions. Based on our experience with this work, it is now a mandatory requirement to have a gender balance in the teams of social workers that the department sends offshore to respond to the needs of Australians impacted by disasters or acts of terrorism. This concept was tested with the first deployment to Bali in 2003 to support families attending the trials of the Bali bombers and it has been found to be an essential response consideration. Factors such as the complexity of family dynamics, the size of families, the nature of the issues, and cultural considerations, have reinforced the need to apply this balance. This principle is applied wherever possible in onshore responses, most recently in Cyclone Yasi in February 2011 particularly in the most severely impacted communities.

In closing, many assumptions can be made about people experiencing the impact of natural disasters and climate change. What we have learned however is not to make assumptions. People and communities react in different ways at different times. It is important to be informed by a body of knowledge and experience but it is equally important to respond to people at the point and pace they choose to engage.

References

Commonwealth of Australia (2008) It's about people: changing perspectives on dryness, a report to government by an expert social panel. Canberra

Department of Human Services (2011) Case studies prepared by social workers involved in these responses. Department of Human Services, Internal document. Canberra

Downey H (2010) Literature review – farm succession planning. Department of Human Services, Internal document. Perth

Hobfall SE, Watson P, Bell CC, Bryant RA, Brymer MJ (2007) Five essential elements of Immediate and mid-term mass trauma intervention: empirical evidence. Psychiatry 70(4, winter):283–315

Pollard G, Oxenham E (2010) Family support drought response team Mt Isa Queensland. Department of Human Services, Internal document. Mt Isa

Roberts L (2010) The resilience tree – drought response team, Ccentral Queensland. Department of Human Services, Internal document. Rockhampton

MIX
Papier aus verantwortungsvollen Quellen
Paper from responsible sources
FSC® C105338

If you have any concerns about our products,
you can contact us on
ProductSafety@springernature.com

In case Publisher is established outside the EU,
the EU authorized representative is:
Springer Nature Customer Service Center GmbH
Europaplatz 3, 69115 Heidelberg, Germany

Printed by Libri Plureos GmbH
in Hamburg, Germany